移动应用系列丛书
丛书主编　倪光南

Android 移动应用开发

肖正兴　主　编
李　斌　张　霞　副主编

中国铁道出版社

2018年·北京

内 容 简 介

本书以多个典型案例阐述移动应用的开发过程,将移动开发的知识点贯穿到案例的实现过程中,最终引领读者进入到移动开发的领域中。

全书共分 5 个项目,涵盖 Android 开发基础知识、基本原理、UI 基本设计、本地存储和网络访问。内容组织上由易到难,层层递进,选取的案例都是典型的企业案例,并简化成教学案例,尽量使初学者能够快速入门。

本书适合作为高等院校理工科相关课程的教材,也可作为开发人员或软件实践者自学和提高的参考书。

图书在版编目(CIP)数据

Android 移动应用开发/肖正兴主编. —北京:中国铁道出版社,2018.5(2018.10重印)
(移动应用系列丛书)
ISBN 978-7-113-24343-2

Ⅰ.①A… Ⅱ.①肖… Ⅲ.①移动终端-应用程序-程序设计 Ⅳ.①TN929.53

中国版本图书馆 CIP 数据核字(2018)第 051064 号

书　　名：Android 移动应用开发
作　　者：肖正兴　主编

策　　划：周海燕	读者热线：(010) 63550836
责任编辑：周海燕　徐盼欣	
封面设计：乔　楚	
责任校对：张玉华	
责任印制：郭向伟	

出版发行：中国铁道出版社(100054,北京市西城区右安门西街 8 号)
网　　址：http://www.tdpress.com/51eds/
印　　刷：三河市宏盛印务有限公司
版　　次：2018 年 5 月第 1 版　2018 年 10 月第 2 次印刷
开　　本:787 mm×1 092 mm　1/16　印张:13　字数:307 千
书　　号：ISBN 978-7-113-24343-2
定　　价：39.00 元

版权所有　侵权必究

凡购买铁道版图书,如有印制质量问题,请与本社教材图书营销部联系调换。电话:(010) 63550836
打击盗版举报电话:市电 (010) 51873659

前言

2007年11月5日，Google公司宣布成立了OHA（Open Handset Alliance，开放手机联盟）包括手机制造商、手机芯片厂商和移动运营商几类，主要包括中国移动、摩托罗拉、英特尔、高通和宏达电，并且开发名为Android的开放源代码的移动系统，到现在为止，使用Android系统的终端设备已超过10亿部，这让它占据了移动应用的半壁江山。本书旨在带领读者进入移动开发的领域。

本书以Eclipse为集成开发环境进行Android应用程序开发，结合笔者近年来在手机软件研发和教学中积累的经验，介绍Android平台移动互联网应用开发的相关知识。

本书共分为5个项目。

项目1介绍Android平台的基本架构和开发环境的配置，并通过编写第一个应用程序Hello World，熟悉Eclipse开发、调试环境的使用。

项目2以一个仿QQ登录界面案例介绍Android基本UI组件的使用，主要包括基本控件的使用和相应的事件处理。

项目3以应用商店案例讲解高级UI适配器组件和相应适配器的使用，主要包括下拉列表控件Spinner、列表视图ListView、网格视图GridView、画廊控件Gallery和选项卡TabHost，还有对应的各种Adapter的配合使用。

项目4以简易的日记本案例介绍Android系统自带的SQLite数据库的使用，实现对本地数据库的增加、删除、修改、查询的操作，并且与ListView控件配合使用。

项目5以天气预报案例介绍Android的HTTP网络通信编程，包括使用HttpClient进行网络访问、数据的解析和多线程的应用。

本书考虑初学者的学习规律和项目的实际开发过程，将实际应用项目抽取、简化成简单、实用的典型教学案例，以项目任务方式使得读者逐步递进学习，并可将现有代码进行适当扩展就能够生成新的应用。

本书由肖正兴任主编，李斌、张霞任副主编。本书的编写得到了深圳职业技术学院等院校的大力支持和帮助，杨勇生为教材案例项目的策划、开发和测试提供了实际的开发项目，中国铁道出版社的编辑为本书的策划和出版提供了宝贵的经验和支持,在此表示衷心感谢。同时,本书在编写过程中,参考了大量的相关资料,吸取了许多同人的宝贵经验,在此一并致谢。

由于编者水平所限,疏漏难免,敬请广大读者提出宝贵意见和建议。教材配套课件、习题答案及源代码均可从 http://www.tdpress.com/51eds/ 下载。

编　者
2018 年 2 月

目录

项目1 移动开发概览 ... 1
- 任务1 配置 Eclipse 移动集成开发环境 ... 7
- 任务2 使用 Eclipse 移动集成开发环境 ... 18
- 任务3 使用 Eclipse 移动集成调试环境 ... 21

项目2 Android 基本 UI 组件——仿 QQ 登录界面 ... 24
- 任务1 实现仿 QQ 登录基本界面 ... 53
- 任务2 实现界面的动态展示 ... 60
- 任务3 实现欢迎界面 ... 66

项目3 Android 高级 UI 组件——应用商店 ... 70
- 任务1 显示商品列表 ... 90
- 任务2 显示分类商品 ... 101
- 任务3 显示商品详情 ... 120
- 任务4 集成应用商店 ... 136

项目4 Android 本地存储——掌上日记本 ... 143
- 任务1 搭建布局文件 ... 151
- 任务2 封装数据操作——适配器 ... 154
- 任务3 搭建主程序 ... 158
- 任务4 编写日记功能 ... 161

项目5 Android 网络通信——天气预报 ... 163
- 任务1 实现天气预报 ... 182
- 任务2 实现天气预报多线程 ... 194

参考文献 ... 202

项目 1　移动开发概览

 项目要点

- Android 环境介绍。
- Eclipse 移动集成开发环境的配置。
- Eclipse 移动集成开发环境的使用。
- 建立和管理 Eclipse 移动开发工程。

本项目主要介绍 Android 平台的基本架构和开发坏境的配置，并通过编写第一个应用程序 HelloWorld，熟悉 Eclipse 开发、调试环境的使用。

 项目简介

本项目主要介绍 Android 的发展历史，搭建 Android 在 Eclipse 下的开发环境，通过第一个应用程序 HelloWorld 介绍 Android 项目的基本运行原理，并在 Eclipse 环境下进行项目开发和调试。

 相关知识

1. Android 概述

1）Android 的概念

当今已经进入移动互联网阶段，这是 50 多年来的第 5 个发展周期。一个技术发展周期一般会持续十年，即"移动互联网"发展周期已经到来。

据摩根士丹利调查：移动互联网的发展速度快于桌面互联网，并且其规模将大得超乎想象，它代表着五大趋势的融合（4G + 社交 + 视频 + 网络电话 + 日新月异的移动装置），在发展速度方面，通过移动装置接入互联网的用户已经超过通过桌面个人计算机接入互联网的用户。到 2020 年，通过各种移动设备连入互联网的设备将达到 100 亿台。

如图 1.1 所示，开放手机联盟（Open Handset Alliance）是 Google 公司于 2007 年 11 月 5 日宣布组建的一个全球性的联盟组织。联盟将会支持 Google 发布的 Android 手机操作系统或者应用软件，共同开发名为 Android 的开放源代码的移动系统。开放手机联盟包括手机制造商、手机芯片厂商和移动运营商等。目前，联盟成员数量已经达到 65 家。

Android 平台具有以下特点。

①Android 平台的源代码完全开放，便于开发人员更清楚地把握实现细节，便于提高开发人员的技术水平，有利于开发出更具差异性的应用。

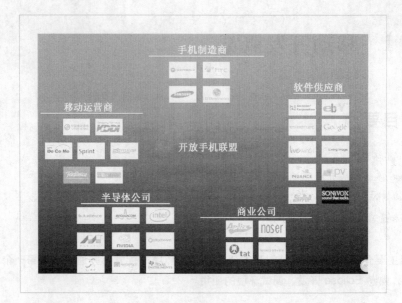

图 1.1 开放手机联盟

②Android 平台采用了对有限内存、电池和 CPU 优化过的虚拟机 Dalvik，有效提升了 Android 的运行速度。

③Android 平台具有良好的盈利模式（3/7 开），产业链条的各方（运营商、制造商、独立软件生产商）都可以获得不错的利益。将移动终端的评价标准从硬件向软件转变，极大地激发了软件开发者的热情。

④Android 的源代码遵循 Apache v2 软件许可，而不是通常的 GPL v2 许可，有利于商业开发。

⑤Android 平台具有强大的 Linux 社区的支持。

Android 曾经是一家创立于旧金山的公司的名字，该公司于 2005 年 8 月被 Google 收购，并从此踏上了飞速发展的道路。它已经发展成为一个平台、一个生态体系。现在，Android 在移动领域得到广泛的应用。相比于 Symbian 的日落西山，以及苹果的封闭、Windows Phone 的前途未卜，Android 无疑代表了当前行业发展的主流趋势：开放平台取代了封闭平台，让参与者均能通过自己的努力而获益。从 2008 年 9 月公布 1.0 版本以来，经过近 10 年的跨越式发展，现在 Android 已经发展到了 8 版本。该版本重新设计通知，以便为管理通知行为和设置提供更轻松和更统一的方式。

随着采用 Android 系统的谷歌手机、平板电脑等产品逐渐扩大市场占有率，Android 平台人才的缺口日益显现。据业内统计，目前国内的 4G 研发人才缺口有三四百万，其中 Android 研发人才缺口至少 30 万。现在不仅手机、平板电脑的企业对 Android 有大量的人才缺口，而且创维、TCL、康佳等彩电企业已经开发了基于 Android 系统的智能电视，Android 已成为 3C 数码电子产品的大潮流。

2）为什么选用 Android

技术流派的发展，一般就像波浪，分为浪峰和浪底，浪峰的时候技术成熟，但是程序员的价值无法得到体现,因为会的人多;浪底的时候,技术新,供参考的资料少,学习门槛高,但是程序员的价值能够得到体现。现在 Android 市场占有量和发展趋势一直向上,人才需求旺盛,而且自主创业点比较多。

(1) 各智能手机平台市场占有率

IDC 调查了各种主流平台市场的占有率,如图 1.2 所示,截至 2014 年,可以看出,作为手机终端的霸主 Symbian 的占有率不断下降;Android 在未来几年的涨幅最大,占有率不断上升;iOS 因为主要针对苹果系统,并且其开发环境是封闭的,一旦其创新性出现断层,必然导致市场占有率的下降。

(2) 主流智能系统的比较

① Symbian——霸主地位已经旁落。Symbian 是 Nokia 手机上的操作系统,在受到各种操作系统的冲击后,倒下了,由于其上面的开发语言为 Visual C++,但是又不同于 PC 上的 C++,导致开发门槛高,而且其转换到智能手机上的步伐偏慢,导致市场占有率不断下降。

Worldwide Converged Mobile Device Operating System Market Shares and 2010-2014 Growth

Operating System	2010 Market Share	2014 Market Share	2014/2010 Change
Symbian	40.1%	32.9%	-18.0%
BlackBerry OS	17.9%	17.3%	-3.5%
Android	16.3%	24.6%	51.2%
iOS	14.7%	10.9%	-25.8%
Windows Mobile	6.8%	9.8%	43.3%
Others	4.2%	4.5%	8.3%
Total	100.0%	100.0%	

IDC分析4年后各智能手机平台市场占有率

图 1.2 智能手机市场占有率

② Android——Google 公司推出,近年势头迅猛。Android 最明智的地方在于两点:联合各大厂商,开放源代码;采用 Java 作为应用开发语言,使得 Java 程序员可以无缝转移到该开发领域,所以虽然其诞生时间不长,但是平台上的应用程序,已经超过了 iOS 上的应用程序。

③ iOS——苹果公司,定位高端客户。iOS 一直针对的是高端客户,但是其需要不断创新,才能保持其高端定位,现在各种智能终端从可使用性上已经和 iOS 无太大差别,并且其开发语言为一种特有的语言:Object C,是在 C 语言基础上加上了面向对象的特性,整个平台是封闭的。

④ Windows Phone——微软+Nokia,全新系统。Windows Phone 是 Windows Mobile 的后续版本,其最大诟病在于与 Windows Mobile 是完全不同的两个东西,没有向下兼容性。

⑤ Meego——Nokia+Intel 共同研发的一种系统。Meego 是一种优秀的手机操作系统,非常流畅,但是 Nokia 并没有打算继续支持其后续版本,所以基本没有前景。

3）Android 的系统架构

图 1.3 所示是 Android 系统架构图，从上而下，分别为应用层、应用框架层、类库层、Android 运行时和 Linux 内核层。

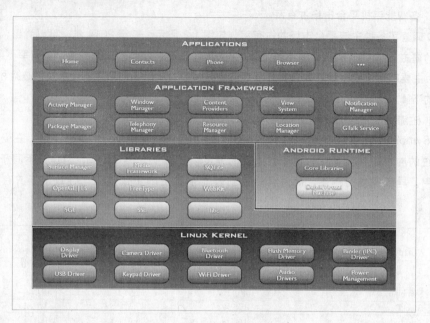

图 1.3　Android 系统架构图

（1）Linux Kernel 层

Android 的核心系统服务依赖于 Linux 2.6 内核，如安全性、内存管理、进程管理、网络协议栈和驱动模型。Linux 内核也同时作为硬件和软件栈之间的抽象层。

（2）Android Runtime 层

Android 包括了一个核心库，该核心库提供了 Java 编程语言核心库的大多数功能。

每个 Android 应用程序都在它自己的进程中运行，都拥有一个独立的 Dalvik 虚拟机实例。Dalvik 被设计成一个设备可以同时高效地运行多个虚拟系统。Dalvik 虚拟机执行（.dex）的 Dalvik 可执行文件，该格式文件针对小内存使用做了优化。同时，虚拟机是基于寄存器的，所有的类都经由 Java 编译器编译，然后通过 SDK 中的 dx 工具转化成 .dex 格式由虚拟机执行。

Dalvik 虚拟机依赖于 Linux 内核的一些功能，比如线程机制和底层内存管理机制。

（3）Libraries 层

Android 包含一些 C/C++ 库，这些库能被 Android 系统中的不同组件使用。它们通过 Android 应用程序框架为开发者提供服务。以下是一些核心库：

①系统 C 库：一个从 BSD 继承来的标准 C 系统函数库（libc），它是专门为基于 embedded Linux 的设备定制的。

②媒体库：基于 PacketVideo OpenCORE。该库支持多种常用的音频、视频格式回放和录制，

同时支持静态图像文件。编码格式包括 MPEG4、H.264、MP3、AAC、AMR、JPG、PNG。

③Surface Manager：对显示子系统进行管理，并且为多个应用程序提供 2D 和 3D 图层的无缝融合。

④LibWebCore：一个 Web 浏览器引擎，支持 Android 浏览器和一个可嵌入的 Web 视图。

⑤SGL：底层的 2D 图形引擎。

⑥3D libraries：基于 OpenGL ES 1.0 APIs 实现，该库可以使用硬件 3D 加速（如果可用）或者使用高度优化的 3D 软加速。

⑦FreeType：位图（bitmap）和矢量（vector）字体显示。

⑧SQLite：一个对于所有应用程序可用、功能强劲的轻型关系型数据库引擎。

（4）Application FrameWork 层

开发人员也可以完全访问核心应用程序所使用的 API 框架。该应用程序的架构设计简化了组件的重用，任何一个应用程序都可以发布它的功能块，任何其他应用程序都可以使用其所发布的功能块（不过需遵循框架的安全性限制）。同样，该应用程序重用机制也使用户可以方便地替换程序组件。

隐藏在每个应用后面的是一系列的服务和系统，其中包括：

①丰富而又可扩展的视图（Views），可以用来构建应用程序，包括列表（lists）、网格（grids）、文本框（text boxes）、按钮（buttons）、甚至可嵌入的 Web 浏览器。

②内容提供器（Content Providers）使得应用程序可以访问另一个应用程序的数据（如联系人数据库），或者共享它们自己的数据。

③资源管理器（Resource Manager）提供非代码资源的访问，如本地字符串、图形和布局文件（layout files）。

④通知管理器（Notification Manager）使得应用程序可以在状态栏中显示自定义的提示信息。

⑤活动管理器（Activity Manager）用来管理应用程序生命周期并提供常用的导航回退功能。

（5）Application 层

Android 会同一系列核心应用程序包一起发布，该应用程序包包括 E-mail 客户端、SMS 短消息程序、日历、地图、浏览器、联系人管理程序等。所有的应用程序都是使用 Java 语言编写的。

4）Android 项目管理

（1）查看项目

在模拟器桌面，单击 Launcher 按钮，打开应用程序管理界面，单击其中的 Android Application 图标，即可启动项目应用程序。

（2）删除项目

①如图 1.4 所示，采用应用程序管理器删除项目。

②如图 1.5 所示，采用 DDMS 删除项目。

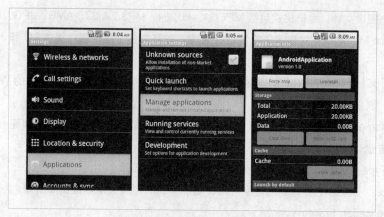

图1.4 应用程序管理器删除项目

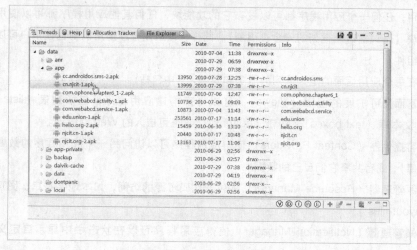

图1.5 DDMS 删除项目

（3）如图1.6所示，在 LogCat 窗口中可以查看项目运行的日志信息。

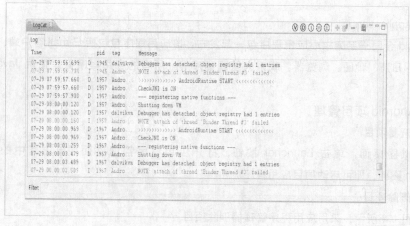

图1.6 LogCat 窗口

2. 应用程序五大组件

Android 应用程序主要包括五大组件，即 Activity、Intent、BroadCastReceiver、Service、Content Provider。

1）Activity

Activity 相当于手机屏幕的一个页面，上面可以放各种 UI 组件。一个应用包含多个 Activity，Android 使用堆栈存放多个 Activity 对象，并自动进行管理。Activity 存在生命周期。

2）Intent

Intent 是一个将要执行的操作的抽象描述（Google）。主要用于在不同 Activity 之间跳转，并且传递数据。主要动作类型包括 Main、PICK、VIEW、EDIT 等，对应的数据以 URI 表示。Intent Filter 用于描述 Activity 能够操作哪些 Intent 对象。

3）BroadCastReceiver

BroadCastReceiver 主要用于对外部事件进行响应。通过 NotificationManager 通知用户事件发生了。通过在 AndroidManifest.xml 中定义，或者通过 context.RegisterReceiver( ) 注册。各种应用还可以通过 Context.sendBroadcast( ) 把自己的 Intent Broadcast 广播给其他应用。

4）Service

Service 是指后台运行的应用程序，如音乐播放器的后台播放。

5）Content Provider

Content Provider 用于在不同应用之间交换数据，一个应用通过实现 Content Provider 的抽象接口，从而把自己的数据暴露给其他应用调用，而其他应用程序无须关心数据存储的具体细节。

任务1　配置 Eclipse 移动集成开发环境

1. 任务说明

Android SDK 所支持的操作系统包括 Windows 7、Mac OS X 10.4.8 or later（x86 only）和 Linux（tested on Linux Ubuntu Dapper Drake）。

搭建 Android 开发环境步骤分为：安装 JDK 6、安装 Eclipse 3.7、安装 ADT 2.2、安装 SDK、配置 AVD。

2. 实现过程

1）安装 JDK 6

首先，到 http://www.oracle.com/technetwork/java/archive-139210.html 下载 Java SE 6，如图1.7所示，然后按照默认配置安装。

2）安装 Eclipse 3.7

①如图1.8所示，进入地址 http://www.eclipse.org/downloads/packages/release/indigo/sr2。

②如图1.9所示，进入 Eclipse IDE for Java Developers 中 Windows 32-bit 版本下载界面，单击 按钮。

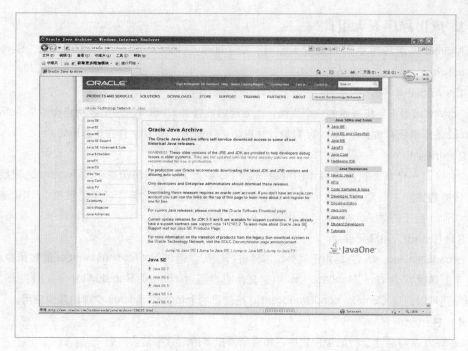

图1.7　Java SE 6 下载界面

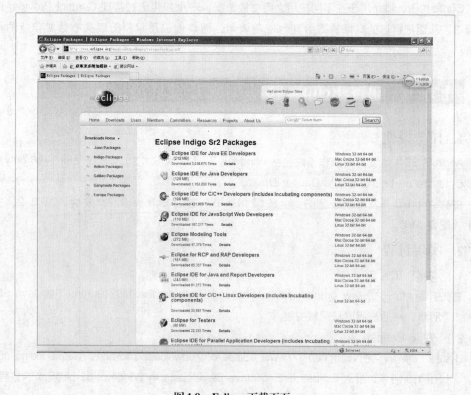

图1.8　Eclipse 下载页面

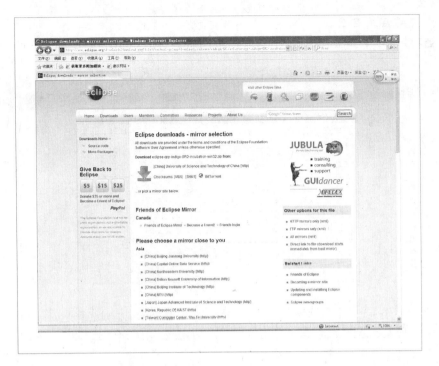

图 1.9 Eclipse IDE

③如图 1.10 所示,单击"保存"按钮。
④解压运行 eclipse.exe。

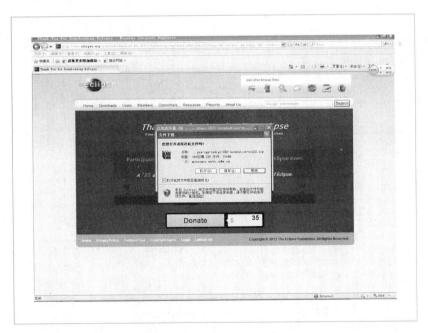

图 1.10 文件下载界面

3）安装 ADT 2.2

①启动 Eclipse，如图 1.11 所示，选取 Help | Install New SoftWare…，单击 Add…按钮。

图 1.11　安装新软件

②如图 1.12 所示，单击 Local…按钮，输入 https：//dl-ssl.google.com/android/eclipse/，单击 OK 按钮。

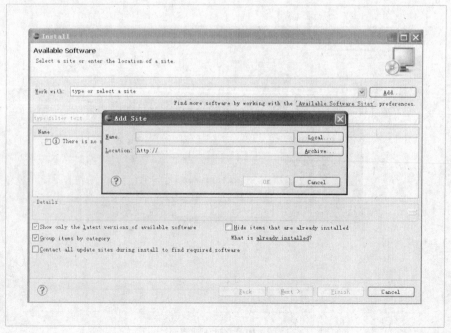

图 1.12　添加站点

③如图 1.13 所示,选择 Developer Tools 复选框,单击 Next 按钮,开始安装。

图 1.13　选择开发工具集

④安装成功后,即可在 Eclipse 中看到 图标。

4)安装 SDK

Software Developer Kit(SDK)的作用是提供 Android 开发所需要的类库的支持,现在最新版本是 7.0,比较流行的版本是 2.2,本书使用 2.2 版本作为教学的 SDK 版本。

第一次运行 Eclipse,要求指定 SDK 的路径,这里分成 SDK 已经存在和 SDK 还没有两种情况。

(1)SDK 已经存在

如图 1.14 所示,SDK 已经安装好了。

①如图 1.15 所示,单击 Browse… 按钮,指定 SDK 的路径。

②单击 Apply… 按钮,如图 1.16 所示,显示出当前 SDK 所支持的 Android 版本。

③单击 OK… 按钮,显示图 1.17 所示的窗口。

④如图 1.18 所示,选择 Use existing SDKs 单选按钮,选择指定的 SDK 的路径,单击 Finish… 按钮。

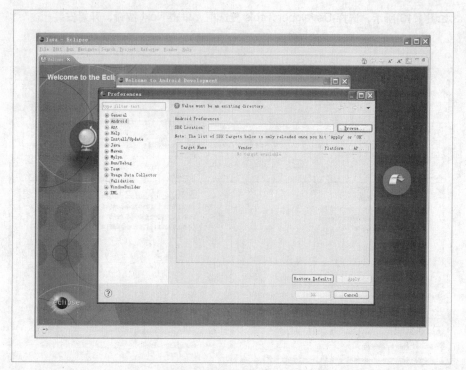

图1.14 选取SDK位置

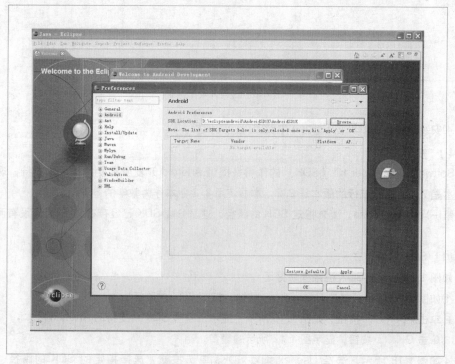

图1.15 指定SDK的路径

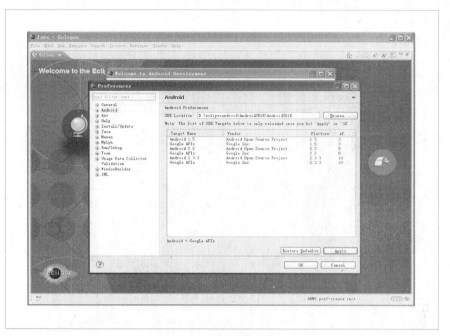

图 1.16　SDK 支持的版本

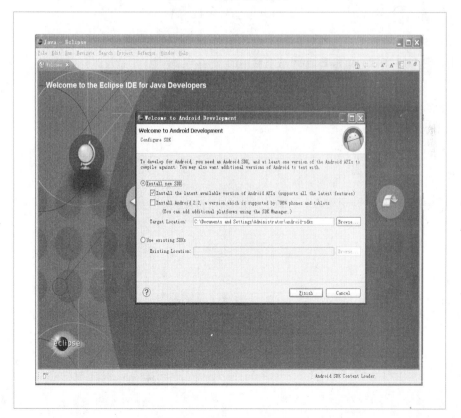

图 1.17　Android 开发欢迎窗口

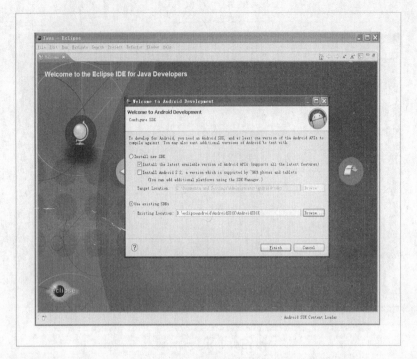

图 1.18　选择存在的 SDK

（2）SDK 不存在

①如图 1.19 所示，单击 Eclipse 任务栏中的 按钮。

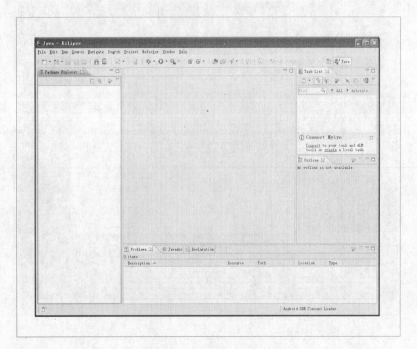

图 1.19　配置 SDK 入口

②如图1.20所示，选择需要安装的SDK版本，这里可以选择2.2版本，其中Tools必须全部选上，否则无法编译和运行。单击Install按钮，等待安装结束，可能需要较长时间，请耐心等待。

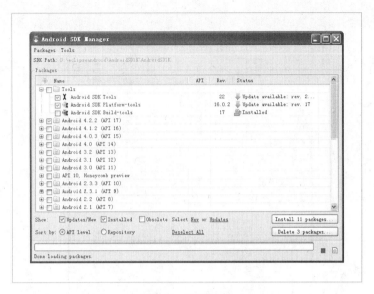

图1.20　选择SDK

5）配置AVD

①如图1.21所示，单击 图标，显示已经配置好的Android虚拟机（AVD），可以根据需要，为不同的SDK配置不同的AVD。

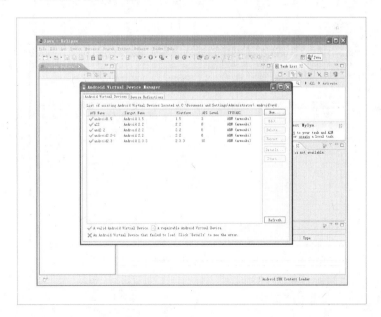

图1.21　配置AVD

②单击 New… 按钮，显示 Create new Android Virtual Device（AVD）对话框，如图 1.22 所示，根据需要选择目标 Device，Target 是 SDK 的版本和相关的内存、SDK 容量。AVD 的作用就是模拟一台实际的手机配置，以便运行相应的 Android 程序。

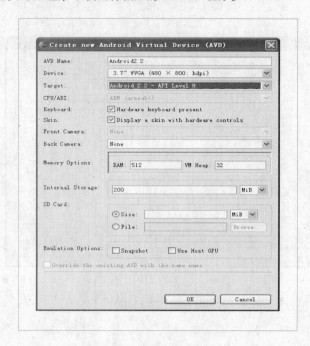

图 1.22 创建新的虚拟机

③单击 OK… 按钮，如图 1.23 所示，可以看到新增配置已经显示在列表中。

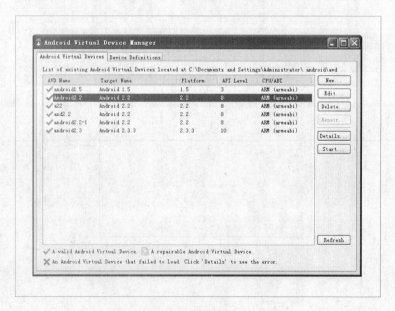

图 1.23 虚拟机列表

④选中一个 AVD 配置,单击 Start... 按钮,如图 1.24 所示,显示 Launch Options 对话框,可以更改启动参数。

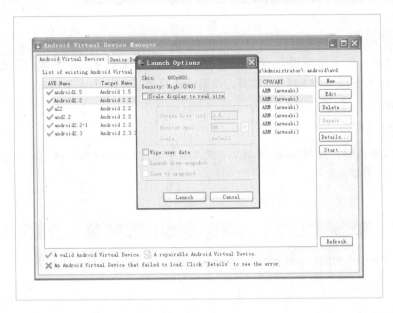

图 1.24　Launch Options 对话框

⑤单击 Launch... 按钮,如图 1.25 所示,可能启动会较慢,请耐心等待。Android 开发环境对内存的需求量比较大,至少 1 GB 内存,2 GB 以上会比较流畅。

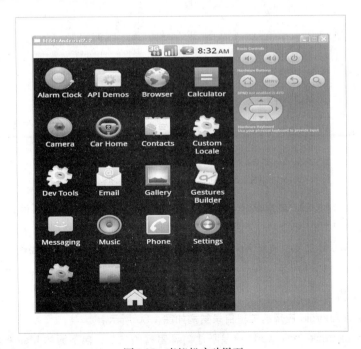

图 1.25　虚拟机启动界面

任务 2　使用 Eclipse 移动集成开发环境

1. 任务说明

在本任务中，将介绍使用 Eclipse 构建 Android 应用程序的基本步骤。首先创建一个 Android 项目。要创建一个 Android 项目，回到 File 菜单，选择 New 项，并在子菜单中选择 Android Application Project 命令。

2. 实现过程

①如图 1.26 所示，在 Application Name 文本框中输入 HelloWorld，在 Theme 下拉列表框中，选择 None 选项。

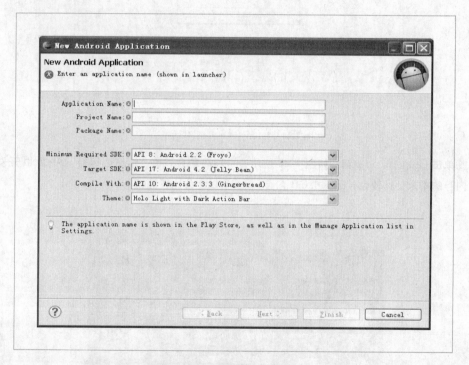

图 1.26　新建 Android 应用向导

②不断单击 Next 按钮，最后单击 Finish 按钮。（注意，这里都使用默认值）最终创建完成的 Android 工程如图 1.27 所示。

③右击 HelloWorld 工程，选择 Run As 中的 Android Application 命令，如图 1.28 所示。

④运行结果如图 1.29 所示。

3. 代码分析

本任务创建的 Android 应用的工程目录结构如图 1.30 所示。

其中，com.example.helloworld 就是刚才向导里起的包名，包下面的 MainActivity.java 则是向导自动生成的 Activity。

图 1.27　Android 工程

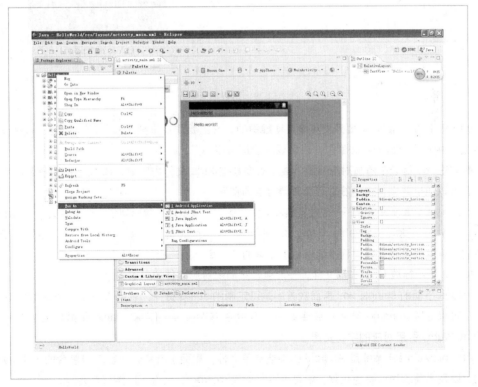

图 1.28　运行程序界面

图1.29　运行结果

gen 包下的 R.java 是由开发环境自动维护的,其主要作用是把工程中需要用到的各种资源文件在 R 中产生相应的静态索引,以便以后快速引用。

每当在 res 目录下添加对应资源,或者改变某些 XML 文件内容时,R.java 的内容都会自动更新,可以通过此文件中的静态资源获得对应的资源。

drawable-＊＊＊文件夹下一般放置图片文件。从 1.6 版本以后就出现了 5 个文件夹:drawable-hdpi、drawable-ldpi、drawable-mdpi、drawable-xdpi、drawable-xxdpi,这些文件夹主要用于存放不同分辨率的目录,以便程序能够根据设备的分辨率选用相应的图片资源文件。

layout 下面存放 UI 的布局文件,一般包含布局及对应控件的组织描述。

values 下面的 strings.xml 里面放置的是可定制的 string 资源,它是一个 key value 类型的键值对,可以通过其 name 引用对应资源。也可以在其 layout 资源及 Manifest 描述文件里通过 @string/name 引用对应的字符资源。

AndroidManifest.xml 是 Android 的描述清单文件,里面主要包含当前应用包含的 Activity 的声明,以及当前应用所有的权限,如是否可以连接 Internet,是否可以拨打电话等。

图 1.30　Android 工程目录结构

任务 3　使用 Eclipse 移动集成调试环境

1. 任务说明

Eclipse Android 开发环境中,有一个非常重要的透视图 DDMS,主要是集成了 Android 调试有用的视图集合的透视图。

2. 实现过程

①单击 Eclipse 任务栏中的 open perspective 按钮　　,显示图 1.31 所示的 Open Perspective 对话框。

②找到 DDMS,单击 OK... 按钮,切换到 DDMS 透视图,如图 1.32 所示。

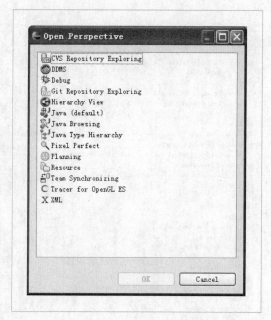

图1.31 Open Perspective 对话框

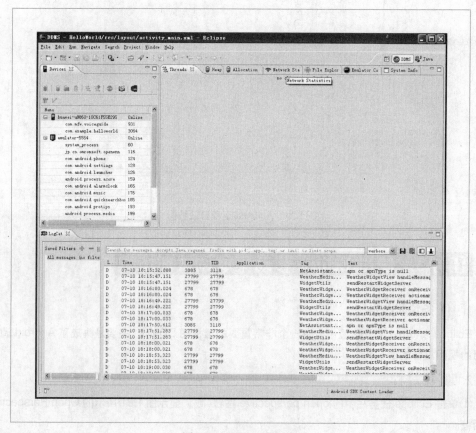

图1.32 DDMS 透视图

DDMS 透视图下，有 Devices 视图、LogCat 视图等。其中，Devices 视图主要列出已经启动的 AVD 列表，包括真机；LogCat 视图主要显示启动 AVD、调试和运行 Android 程序的调试信息，是非常重要的调试视图。

 实战演练

1. 安装和配置 Eclipse 开发环境。
2. 编写程序，在手机上打印输出"欢迎参加 Android 开发，加油！"。
3. 下载一个 .apk 文件并安装到模拟器。

项目 2 Android 基本 UI 组件——仿 QQ 登录界面

 项目要点

- 了解图形用户界面的基本概念。
- 理解高级图形用户界面的组成。
- 熟练使用基本的 Android 控件。
- 使用事件与各种窗口元素。

项目 1 初步介绍了 Android 平台的基本架构、开发环境，并编写了第一个应用程序 HelloWorld。本项目将介绍 Android 的基本 UI 组件的使用。

图形用户界面（Graphic User Interface，GUI）是用户与应用程序交互的接口。一个应用程序界面设计的好坏，将直接影响用户使用程序的体验。风格漂亮、使用便捷、设计合理的程序界面是吸引用户的一个重要因素。在传统 PC 中，应用程序的图形界面功能强大，程序员可根据需要，设计出风格各异、千变万化的界面。但对于手机等手持移动设备来说，由于受屏幕小、计算能力弱、输入不便等因素制约，决定了 Android 平台的图形用户界面，无论是可用的窗口、组件类型、使用的方法，还是编程的方式，都与传统 PC 的图形用户界面有很大区别。

 项目简介

本项目综合运用布局、基本 UI 组件、Activity 及其跳转等相关知识，搭建了一个仿 QQ 登录的界面，并实现基本功能，项目的运行效果如图 2.1 所示。

图 2.1 仿 QQ 登录项目的运行效果

相关知识

Android 设备的用户界面模型非常简单。在传统 PC 中，系统经常同时打开多个应用程序，每个应用程序能包含多个窗口。用户通过点击鼠标，可以在各个窗口之间切换。而对于 Android 设备来说，屏幕每次只能显示一个"窗口"。这意味着，如果设备中同时有多个应用程序在运行，在某一时刻，只有一个 Android 程序能够使用屏幕，显示其窗口。并且，一个 Android 程序如果由多个窗口组成，一次也只能显示一个窗口。

在 Android 平台上，用户界面通过 ViewGroup 或 View 类来显示。ViewGroup 和 View 类是 Android 平台上最基本的用户界面表达单元。可以通过程序直接调用的方法调用描绘用户界面。既可以将屏幕上显示的界面元素与构成应用程序主体的程序逻辑混合在一起编写，通过代码创建界面，也可以将界面显示与程序逻辑分离，使用 XML 文档来描述和生成界面。

用户界面组件的继承关系如图 2.2 所示。

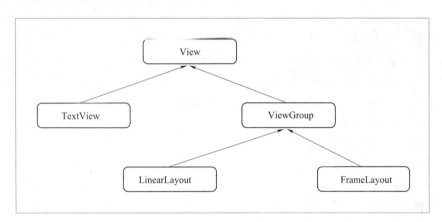

图 2.2 用户界面组件的继承关系

1. 认识 Activity

为了在应用程序中创建用户界面，离不开 Activity。一个 Activity 代表一个显示给用户的屏幕，也就是前面说的"窗口"。通常，应用程序需要至少一个 Activity 作为主屏幕，同时还需要其他 Activity 作为补充。

2. 创建 Activity

创建 Activity 比较简单，只需继承 Activity 类即可，下面的代码展示了如何创建一个新的 Activity。

import android. os. Bundle；
import android. app. Activity；
import android. view. Menu；
public class MainActivity **extends** Activity {
　@ Override

```
    protected void onCreate(Bundle savedInstanceState){
        super.onCreate(savedInstanceState);
    }
}
```

这样的 Activity 将显示一个没有任何内容的空窗口，为了往这个空窗口中加入希望的内容，可以将各种视图资源（View 或者 ViewGroup）填充到这个空窗口中。例如若希望将一个 TextView 加入进来，则可以在 onCreate( )方法中调用 setContentView( )方法，代码如下所示：

```
@Override
protected void onCreate(Bundle savedInstanceState){
    super.onCreate(savedInstanceState);
    TextView m = new TextView(this);
    setContentView(m);
}
```

3. Activity 的生命周期

Activity 继承自 ApplicationContext，可通过重写以下方法：

```
public class Activity extends ApplicationContext {
        protected void onCreate(Bundle savedInstanceState);
        protected void onStart();
        protected void onRestart();
        protected void onResume();
        protected void onPause();
        protected void onStop();
        protected void onDestroy();
}
```

图 2.3 所示为 Activity 生命周期结构图。Activity 启动后首先进入 onCreate( )方法，顾名思义这个方法在 Activity 开始创建时调用，可以在其中定义一些初始化操作。接下来是 onStart( )方法，这个方法在 Activity 开始被执行时调用，它紧随 onCreate( )方法之后调用。接着是 onResume( )方法，这个方法是在该 Activity 获得用户输入焦点时被调用，当这个方法调用后 Activity 开始真正运行。在 Activity 正在运行时用户激活了另一个 Activity，这时将调用第一个 Activity 的 onPause( )方法，可以理解为第一个 Activity 被暂停了，这时如果系统的内存不够用（手机内存不够用的情况经常发生），第一个 Activity 的进程可能被杀死（何时被杀死是由系统决定的），当下次再运行第一个 Activity 时就需要重新创建这个 Activity，那就又需要调用 onCreate( )方法。如果在这个 Activity 没有被杀死的情况下重新调用第一个 Activity，就会直接调用它的 onResume( )方法后开始运行。如果第一个 Activity 很久都没有得到再次运行的机会，就会调用 onStop( )方法被停止，这时如果 Activity 又获得用户输入焦点，就会调用 onRestart( )方法，重新开始执行这个 Activity，或者被系统杀死，否则调用 onDestroy( )方法销毁 Activity。

项目 *2*　Android 基本 UI 组件——仿 QQ 登录界面

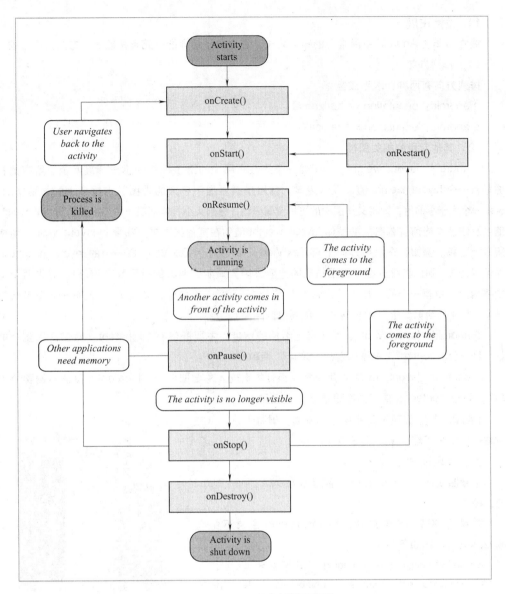

图 2.3　Activity 生命周期结构图

4．布局

在一个 Activity 中通常会放置多个 UI 组件。那么，如何将这些 UI 组件放置在屏幕的合适位置？需要容器来存放这些控件并控制它们的位置排列，就像 HTML 中 div、table 一样，Android 布局也起到同样的作用。

Android 布局主要有以下几种：线性布局（LinearLayout）、相对布局（RelativeLayout）、表格布局（TableLayout）、帧布局（FrameLayout）和绝对布局（AbsoluteLayout）。其中，AbsoluteLayout 是通过指定控件的 x/y 坐标来定位的，不太灵活，所以已经不再推荐使用。下面对前四种布局分别进行介绍。

1）线性布局

线性布局是 Android 中最常用的布局之一，它将自己包含的子元素按照一个方向进行排列。

（1）排列方向

排列方向有两种：水平或竖直。

①android：orientation = " horizontal"

②android：orientation = " vertical"

（2）其他常用的相关属性

①android：layout_weight：用于给一个线性布局中的诸多视图的重要程度赋值。所有的视图都有一个 layout_weight 值，默认为零，意思是需要显示多大的视图就占据多大的屏幕空间。若赋一个大于零的值，则将父视图中的可用空间分割，分割大小具体取决于每个视图的 layout_weight 值以及该值在当前屏幕布局的整体 layout_weight 值和在其他视图屏幕布局的 layout_weight 值中所占的比率。例如，在水平方向上有 TextView。如果两个 TextView 每一个的 layout_weight 值都设置为 1，则两者平分在父视图布局的宽度（因为声明这两者的重要程度相等）。如果两个文本编辑元素中第一个的 layout_weight 值设置为 1，而第二个的值设置为 2，则空间的 2/3 分给第一个，1/3 分给第二个（数值越小，重要程度越高）。

②android：layout_width：用于指定组件的宽度。主要取值有 fill_parent（宽度占满整个屏幕）和 wrap_content（根据内容设置宽度）两种。

③android：layout_height：用于指定组件的宽度。主要取值有 fill_parent（高度占满整个屏幕）、wrap_content（根据内容设置高度）两种。

【例2-1】实现图2.4所示的界面。屏幕上方，从左至右依次排列了红、绿、蓝三种颜色块。

（1）实现过程

①参照项目1，在 Eclipse 中创建名为 ex02_linearlayout 的工程。

②编写 XML 布局文件，在 res/layout 目录下编写 ex02.xml，代码如下所示。

<? xml version = " 1.0" encoding = " *utf-8* " ? >

< LinearLayout xmlns：android = " *http://schemas.android.com/apk/res/android*"

    android：layout_width = " *fill_parent*"

    android：layout_height = " *fill_parent*"

    android：orientation = " *horizontal*" >

      < TextView

        android：id = " *@ + id/textView*1"

        android：layout_width = " *fill_parent*"

        android：layout_height = " *fill_parent*"

        android：layout_weight = " 1"

        android：background = " *#FF*0000" / >

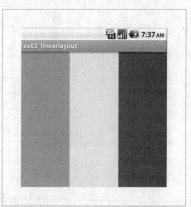

图2.4　ex02_linearlayout 程序

```xml
<TextView
    android:id = "@+id/textView2"
    android:layout_width = "fill_parent"
    android:layout_height = "fill_parent"
    android:layout_weight = "1"
    android:background = "#00FF00" />
<TextView
    android:id = "@+id/textView3"
    android:layout_width = "fill_parent"
    android:layout_height = "fill_parent"
    android:layout_weight = "1"
    android:background = "#0000FF" />
</LinearLayout>
```

③修改 MainActivity.java 的代码，设置 Activity 的界面为所编写的 XML 文件，代码如下：

```java
package com.example.ex02_linearlayout;
import android.os.Bundle;
import android.app.Activity;
import android.view.Menu;
public class MainActivity extends Activity {
    @Override
    protected void onCreate(Bundle savedInstanceState) {
        super.onCreate(savedInstanceState);
        setContentView(R.layout.linearlayout);
    }
}
```

（2）代码分析

例 2-1 中最外层的布局为水平的线性布局，依次向该布局中加入 TextView，分别设置 TextView 的背景色。结构如图 2.5 所示。

（3）功能扩展

修改例 2-1 的代码，实现图 2.6 所示界面。可以首先将整个界面看成两部分，垂直排列，分别为水平方向和垂直方向的线性布局，通过指定 layout_weight，让这两个布局各占一半。最后依次向内部的两个线性布局中加入 TextView。

2）相对布局

相对布局就是在确定界面元素的位置时，使用相对位置，可以相对其他元素，也可以相对这个布局。一个视图可以指定相对于它的兄弟视图的位置（如在给定视图的左边或者下面）或相对于 RelativeLayout 的特定区域的位置（如底部对齐或中间偏左）。使用相对布局可以有效消除嵌套布局。如果使用了多个嵌套的 LinearLayout 视图组，可以考虑使用一个相对布局。

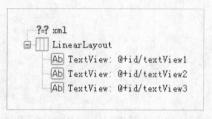
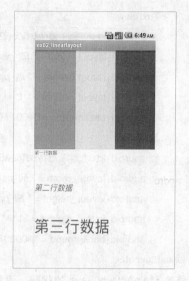

图 2.5　ex02_linearlayout 的布局结构　　　图 2.6　修改后的界面效果

相对布局相关属性如下：

- android：layout_above：将该控件的底部置于给定 ID 的控件之上；
- android：layout_below：将该控件的顶部置于给定 ID 的控件之下；
- android：layout_toLeftOf：将该控件的右边缘和给定 ID 的控件的左边缘对齐；
- android：layout_toRightOf：将该控件的左边缘和给定 ID 的控件的右边缘对齐；
- android：layout_alignBaseline：将该控件的 baseline 和给定 ID 的控件的 baseline 对齐；
- android：layout_alignBottom：将该控件的底部边缘与给定 ID 控件的底部边缘对齐；
- android：layout_alignLeft：将该控件的左边缘与给定 ID 控件的左边缘对齐；
- android：layout_alignRight：将该控件的右边缘与给定 ID 控件的右边缘对齐；
- android：layout_alignTop：将该控件的顶部边缘与给定 ID 控件的顶部边缘对齐；
- android：layout_alignParentBottom：如果该值为 true，则将该控件的底部和父控件的底部对齐；
- android：layout_alignParentLeft：如果该值为 true，则将该控件的左边缘和父控件的左边缘对齐；
- android：layout_alignParentRight：如果该值为 true，则将该控件的右边缘和父控件的右边缘对齐；
- android：layout_alignParentTop：如果该值为 true，则将该控件的顶部和父控件的顶部对齐；
- android：layout_centerHorizontal：如果该值为 true，则该控件将被置于水平方向的中央；
- android：layout_centerInParent：如果该值为 true，则该控件将被置于父控件中，水平方向和垂直方向都居中；
- android：layout_centerVertical：如果该值为 true，则该控件将被置于垂直方向的中央。

【例2-2】 实现图2.7所示的界面。

(1) 实现过程

①参照项目1,在Eclipse中创建名为ex02_relativelayout的工程。

②编写XML布局文件,在res/layout目录下编写ex02_relativelayout.xml,代码如下所示。

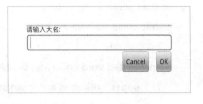

图2.7 相对布局

```xml
<?xml version="1.0" encoding="utf-8"?>
<RelativeLayout xmlns:android="http://schemas.android.com/apk/res/android"
    android:layout_width="fill_parent"
    android:layout_height="fill_parent" >
    <TextView
        android:id="@+id/label"
        android:layout_width="fill_parent"
        android:layout_height="wrap_content"
        android:text="请输入大名:" />
    <EditText
        android:id="@+id/entry"
        android:layout_width="fill_parent"
        android:layout_height="wrap_content"
        android:layout_below="@id/label" />
    <Button
        android:id="@+id/ok"
        android:layout_width="wrap_content"
        android:layout_height="wrap_content"
        android:layout_below="@id/entry"
        android:layout_alignParentRight="true"
        android:layout_marginLeft="10dip"
        android:text="OK" />
    <Button
        android:layout_width="wrap_content"
        android:layout_height="wrap_content"
        android:layout_toLeftOf="@id/ok"
        android:layout_alignTop="@id/ok"
        android:text="Cancel" />
</RelativeLayout>
```

③修改MainActivity.java的代码,设置Activity的界面为所编写的XML文件,代码如下:

```
package com.example.ex02_relativelayout;
import android.os.Bundle;
import android.app.Activity;
```

```
import android. view. Menu;
public class MainActivity extends Activity {
    @Override
    protected void onCreate( Bundle savedInstanceState) {
        super. onCreate( savedInstanceState) ;
        setContentView( R. layout. relativelayout) ;
    }
}
```

（2）代码分析

例 2-2 中最外层的布局为相对布局，首先放置 TextView 的位置，然后通过设置 EditText 组件的 android：layout_below = " @id/label"，来确定 EditText 组件的位置。通过设置 OK 按钮的 android：layout_alignParentRight = "true" 属性，指定该按钮靠容器右对齐，再设置 android：layout_marginLeft = "10 dip" 指定其与边的间距为 10 dip，从而确定了该按钮的位置。最后通过设置与 OK 按钮的相对位置，从而确定 Cancel 按钮的位置。ex02_relativelayout 的布局结构如图 2.8 所示。

（3）功能扩展

修改例 2-2 的布局文件代码，实现图 2.9 所示的界面。要求两个按钮相对于屏幕边缘的距离相等，即两个按钮位于屏幕中央。

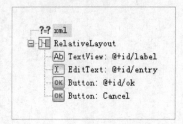 

图 2.8 ex02_relativelayout 的布局结构   图 2.9 修改后的界面效果

3）表格布局

所谓表格布局，就是以表格形式显示并放置 UI 组件，即以行和列标识一个组件的位置。其实 Android 的表格布局与 HTML 中的表格布局非常类似，TableRow 就像 HTML 表格的 < tr > 标记。表格布局相关属性如下：

①android：shrinkColumns，对应的方法为 setShrinkAllColumns（boolean），作用为设置表格的列是否收缩（列编号从 0 开始，下同），多列用逗号隔开（下同），如 android：shrinkColumns = " 0，1，2"，即表格的第 1、2、3 列的内容是收缩的以适合屏幕，不会挤出屏幕。

②android：collapseColumns，对应的方法为 setColumnCollapsed（int，boolean），作用为设置表格的列是否隐藏。

③android：stretchColumns，对应的方法为 setStretchAllColumns（boolean），作用为设置表格的列是否拉伸。其效果就是在其他 column 可以完整显示时，该 column 就会伸展，占最多空间。

④android:layout_column,可以设置 index 值实现跳开某些单元格。

【例2-3】 实现图 2.10 所示的界面。

(1) 实现过程

①参照项目1,在 Eclipse 中创建名为 ex02_tablelayout 的工程。

②编写 XML 布局文件,在 res/layout 目录下编写 ex02_tablelayout.xml,代码如下所示。

图 2.10 表格布局

```xml
<?xml version="1.0" encoding="utf-8"?>
<TableLayout xmlns:android="http://schemas.android.com/apk/res/android"
    android:layout_width="fill_parent"
    android:layout_height="fill_parent"
    android:stretchColumns="0,1,2"
    >
    <TableRow android:background="#0000FF">
        <TextView android:text="第一行第一列"
            android:padding="3dip"
            android:gravity="center"
            android:textColor="#FFFFFF"/>
        <TextView android:text="第一行第二列"
            android:padding="3dip"
            android:gravity="center"
            android:textColor="#FFFFFF"/>
        <TextView android:text="第一行第三列"
            android:padding="3dip"
            android:gravity="center"
            android:textColor="#FFFFFF"/>
    </TableRow>
    <TableRow android:background="#FF00FF">
        <TextView android:text="第二行第一列"
            android:padding="3dip"
            android:gravity="center"
            android:textColor="#00FF00"/>
        <TextView android:text="第二行第二列"
            android:padding="3dip"
            android:gravity="center"
            android:textColor="#00FF00"/>
        <TextView android:text="第二行第三列"
            android:padding="3dip"
            android:gravity="center"
```

android:textColor="#00FF00"/>

</TableRow>

</TableLayout>

③修改 MainActivity.java 的代码，设置 Activity 的界面为所编写的 XML 文件，代码如下：

```
package com.example.ex02_tablelayout;
import android.os.Bundle;
import android.app.Activity;
import android.view.Menu;
public class MainActivity extends Activity {
    @Override
    protected void onCreate(Bundle savedInstanceState) {
        super.onCreate(savedInstanceState);
        setContentView(R.layout.ex02_tablelayout);
    }
}
```

（2）代码分析

例 2-3 中最外层的布局为表格布局，通过放入两个 TableRow，构建两行，然后在每行中依次加入 TextView。ex02_tablelayout 的布局结构如图 2.11 所示。

（3）功能扩展

修改例 2-3 的布局文件代码，实现图 2.12 所示的界面效果。

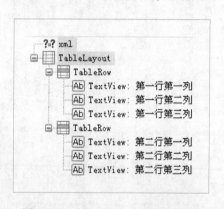

图 2.11　ex02_tablelayout 的布局结构

图 2.12　修改后的界面效果

4）帧布局

所谓帧布局，就是在它内部的元素是一层一层地叠加在一起的。层的概念和 Photoshop 或者 Flash 中层的概念是一致的。如果最上层的元素是不透明的，并且比下面的元素尺寸要大，那么将看不到下面的元素，只能看到顶层元素。这些层的顺序是：最新声明的放到最前面。可以这样理解：Android 按文件的书写顺序来组织这个布局，先声明的放在第一层，再声明的放到第二层，最后声明的放在最顶层。这种布局相对比较简单，这里就不做详细说明了。

5. Android 基本的 UI 组件概述

1） View 与 ViewGroup

View 是 Android 中可视化组件的基类，主要提供了组件的绘制和事件处理的方法。而可视化组件则是指重新实现了 View 的绘制和事件处理方法并最终与用户交互的对象。如 TextView、Button 等。

ViewGroup 类继承自 View 类，其最大的特点就是可以有子组件，而且可以嵌套。如各种布局类。

2） TextView 与 EditText

TextView 继承自 View 类，显示文本并提供允许编辑的选项。TextView 是完整的文本编辑器，只是基类设置为不可编辑，所以通常用来显示文本信息。

EditText 继承自 TextView，只是对 TextView 进行了少量变更，以使其可以编辑。

3） Button

Button 代表按钮小部件。用户通过按下按钮，或者单击按钮来执行一个动作。Button 类继承自 TextView。

4） ImageView 与 ImageButton

ImageView 显示任意图像，例如图标。ImageView 类可以加载各种来源的图片（如资源或图片库），需要计算图像的尺寸，以便它可以在其他布局中使用，并提供缩放和着色（渲染）等各种显示选项。

ImageButton 继承自 ImageView。默认情况下，ImageButton 看起来像一个普通的按钮，拥有标准的背景色，并在不同状态时变更颜色。按钮上的图片可用通过 XML 布局文件的 <ImageButton> XML 元素的 android：src 属性或者代码中的 setImageResource（int）方法指定。

要移除标准按钮背景图像，可以定义自己的背景图片或设置背景为透明。

为了表示不同的按钮状态（得到焦点、被选中等），可以为每种状态定义不同的图片。例如，默认状态为蓝色图片，获得焦点时显示橙色图片，按下时显示黄色图片。使用 XML 布局文件的可绘制对象 selector 可以简单地实现该功能。

5） CheckBox 与 RadioButton

复选框 CheckBox 是包含选中和未选中两种状态的特殊的双状态按钮。

6. Android 基本的 UI 组件应用

1） Button 和 ImageButton

Button 是各种 UI 中最常用的控件之一，它也是 Android 开发中最受欢迎的控件之一，用户可以通过触摸它来触发一系列事件。Button 类继承自 TextView 类，因此，TextView 类的一些属性同样适用于 Button。主要属性如下：

①android：id：给 Button 设置唯一的名字。通常写法为 android：id = " @ + id/btnOK"。其中，"+"表示通过它来生成静态资源，如果没有"+"，表示使用的是指定位置的静态资源。一般为控件赋 ID 时，在使用 + 这个方法保存 XML 后，可以发现 R.java 中已经有一个内部类 id，这个 id 类有一个静态字段叫 btnOK，它的具体值不必关心，可以在代码中通过 R.id.btnOK 来获得它的值。

②android：text：设置按钮显示的文字，如果 android：layout_width 用了 wrap_content 属性，这个文字的长度将会隐式地决定按钮的宽度。该属性可以直接赋值，如 android：text = "确定"；也可以利用资源文件进行设置，如首先在 res\\values\\strings.xml 中写入 <string name=" btnText" >确定</string>，然后设置 android：text = " @string/btnText"，这样系统就会把 btnText 所对应的值作为 Button 的值。如果要把"确定"改成"取消"，只需要改变 strings.xml 中的值即可，不需要改动任何 Java 代码。这在需要将项目移植为其他语言版本时非常有用，如要将软件改为英文版本，可以将"确定"改成 OK，而不需要重新编译。

③android：layout_width：设置按钮的宽度，必须设置。

④android：layout_height：设置按钮的高度，必须设置。

ImageButton 继承自 ImageView，与 Button 非常类似，只是没有 android：text 属性，而是使用 src 属性设置 ImageButton 控件上要显示的图片。

【例 2 - 4】 实现图 2.13 所示的界面。运行程序后，单击左边的按钮，屏幕显示"您点击了 Button 按钮"；单击右边的 ImageButton，屏幕显示"您点击了 ImageButton 按钮"。

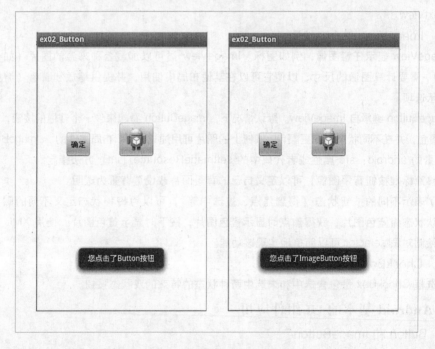

图 2.13　ex02_Button 程序

(1) 实现过程

①参照项目1，在 Eclipse 中创建名为 ex02_Button 的工程。

②打开 res\values\strings.xml 文件，写入 <string name=" btnText" >确定</string>。

③修改布局文件 activity_main.xml，代码如下所示。

< RelativeLayout xmlns：android = " http://schemas.android.com/apk/res/android"
　　xmlns：tools = " http://schemas.android.com/tools"

```xml
android:layout_width = "match_parent"
android:layout_height = "match_parent"
android:paddingBottom = "@dimen/activity_vertical_margin"
android:paddingLeft = "@dimen/activity_horizontal_margin"
android:paddingRight = "@dimen/activity_horizontal_margin"
android:paddingTop = "@dimen/activity_vertical_margin"
tools:context = ".MainActivity" >
<Button
    android:id = "@+id/btnOK"
    android:layout_width = "wrap_content"
    android:layout_height = "wrap_content"
    android:layout_alignParentLeft = "true"
    android:layout_alignParentTop = "true"
    android:layout_marginLeft = "37dp"
    android:layout_marginTop = "141dp"
    android:text = "@string/btnText" />
<ImageButton
    android:id = "@+id/imgBtn"
    android:layout_width = "wrap_content"
    android:layout_height = "wrap_content"
    android:layout_alignBottom = "@+id/btnOK"
    android:layout_marginLeft = "50dp"
    android:layout_toRightOf = "@+id/btnOK"
    android:src = "@drawable/ic_launcher" />
</RelativeLayout>
```

④在 MainActivity.java 中，修改代码，如下所示。

```java
import android.os.Bundle;
import android.app.Activity;
import android.view.Menu;
import android.view.View;
import android.view.View.OnClickListener;
import android.widget.Button;
import android.widget.ImageButton;
import android.widget.Toast;
public class MainActivity extends Activity {
    Button btnOK;
    ImageButton imgBtn;
    @Override
    protected void onCreate(Bundle savedInstanceState) {
        super.onCreate(savedInstanceState);
```

```
        setContentView( R. layout. activity_main);
        btnOK = ( Button ) findViewById( R. id. btnOK);
        imgBtn = ( ImageButton ) findViewById( R. id. imgBtn);
        OnClickListener listener = new OnClickListener( ) {
            @Override
            public void onClick( View v) {
                // TODO Auto - generated method stub
                if( v = = btnOK) {
                    Toast. makeText( MainActivity. this, "您点击了 Button 按钮", Toast. LENGTH_LONG). show( );
                }
                if( v = = imgBtn) {
                    Toast. makeText( MainActivity. this, "您点击了 ImageButton 按钮", Toast. LENGTH_LONG). show( );
                }
            }
        };
        btnOK. setOnClickListener( listener);
        imgBtn. setOnClickListener( listener);
    }
}
```

(2) 代码分析

①在 Activity 中获取到 Button 的实例 btnOK = (Button) findViewById (R. id. btnOK)。通过 findViewById,可以获取指定 Id 的组件实例。但是 Id 值是一串数字,如 public static final int btnOK =0x7f080000;,非常难以记住,因此,合适的做法是利用 R. id 来获取指定的名称,而这个名称又唯一对应了控件 ID。如果系统不能识别 Button 类,则需要导入 Android 的 Button 所在的类包 import android. widget. Button。

通常为了方便起见,可以定义成员变量 Button btnOK,用来存储在 onCreate 中利用 findViewById 找到的 Button 实例,这样,在后面就不需要再寻找了。

②要让按钮响应用户单击的事件,需要为按钮设置监听器。下面的代码定义了 ClickListener 监听器:

```
OnClickListener listener = new OnClickListener( ) {
    @Override
    public void onClick( View v) {
        // TODO Auto - generated method stub
        if( v = = btnOK) {
            //单击 Button
        }
        if( v = = imgBtn) {
```

//单击 ImageButton
            }
        }
    };

③监听器。监听器是个抽象类，它包含了一个事件触发时系统会去调用的方法。在子类中，根据项目的需要重写这个方法。

派生后的监听器需要绑定到按钮上，就像一个耳机可以发出声音，但不去戴它则听不到它发出的声音。一般情况是这个按钮可能需要这个监听器，而另外一个按钮需要另外一个监听器，每个监听器各司其职。但功能相似时，也可以多个按钮共同绑定一个监听器。

各种控件都有常用的事件，如单击按钮、拖动一个滚动条、切换一个 ListView 的选项等。绑定监听器的函数命名规则是 setOn＊＊＊Listener。

例如，btnOK.setOnClickListener（listener）：将监听器对象 listener 设置到 btnOK 上；imgBtn.setOnClickListener（listener）将监听器对象 listener 设置到 imgBtn 上。因为这两个组件共用一个监听器，因此，在处理程序中，需要通过 if（v＝＝btnOK）来判断是哪个对象触发的事件。

（3） 功能扩展

为例 2-4 添加动态生成按钮的功能，当用户单击 imgBtn 后，程序自动向界面中添加一个 Button 按钮，如图 2.14 所示。

2） TextView 和 EditText

（1） TextView

TextView 主要用来显示文本。

①常用的属性如下：

• android：textColor：用于设置字体的颜色，如"#ff8c00"；

• android：textStyle：用于设置字体的样式，如：bold（粗体）、italic（斜体）等；

• android：textSize：用于设置文字的大小，如"20sp"；

• android：textAlign：用于设置文字的排列方式，如"center"；

• android：background：用于设置组件的背景色，如"#FF0000"；

• android：layout_weight：与 LinearLayout 中的意义相同。

图 2.14 动态添加按钮效果

②主要方法如下：

• getText( )获取 TextView 中文本的内容；

• setText( )设置 TextView 中文本的内容。

（2） EditText

EditText 继承自 TextView，因此，TextView 有的属性在 EditText 中也大都存在。EditText

提供了用户输入信息的接口，是实现人机交互的重要组件。

①主要属性如下：
- android：layout_gravity：设置组件显示的位置，默认为 top，还有 bottom、center_vertical 等；
- android：hint：当没有输入时，设置显示在组件上的提示信息；
- android：numeric：设置 EditText 只能输入数字，数字类型一共有三种，分别为 integer（正整数）、signed（带符号整数）和 decimal（浮点数）；
- android：singleLine：设置是否为单行输入，一旦设置为 true，则文字不会自动换行。
- android：password：设置只能密码输入功能；
- android：phoneNumber：设置输入电话号码；
- android：textAlign：设置文本对齐方式；
- android：autoText：自动拼写帮助；
- android：editable：是否可编辑；
- android：textColorHighlight：被选中文字的底色，默认为蓝色；
- android：textColorHint：设置提示信息文字的颜色，默认为灰色；

②主要方法如下：
- getText( )：获取 EditText 中文本的内容；
- setText( )：设置 EditText 中文本的内容。

【例 2-5】 实现登录功能，当用户输入用户名和密码后，单击"登录"按钮，程序验证用户名和密码不为空，则显示"登录成功"提示信息，否则，出现"用户名和密码不能为空"的信息。当用户单击"重置"按钮，则清空两个 EditText 中输入的信息，如图 2.15 所示。

(1) 实现过程

①参照项目 1，在 Eclipse 中创建名为 ex02_Login 的工程。

②打开 res \ values \ strings.xml 文件，写入如下内容：

< string name = " *txtUser*" >用户名：</string >

< string name = " *txtPwd*" >密码：</string >

< string name = " *btnLogin*" >登录</string >

< string name = " *btnReset*" >重置</string >

③修改布局文件 activity_main.xml，代码如下所示：

< RelativeLayout xmlns：android = " *http：//schemas.android.com/apk/res/android*"

  xmlns：tools = " *http：//schemas.android.com/tools*"

  android：layout_width = " *match_parent*"

  android：layout_height = " *match_parent*"

  android：paddingBottom = " *@dimen/activity_vertical_margin*"

  android：paddingLeft = " *@dimen/activity_horizontal_margin*"

  android：paddingRight = " *@dimen/activity_horizontal_margin*"

图 2.15 登录程序

```xml
        android:paddingTop = "@dimen/activity_vertical_margin"
    tools:context = ".MainActivity" >
    <TextView
        android:id = "@+id/textView1"
        android:layout_width = "wrap_content"
        android:layout_height = "wrap_content"
        android:layout_alignParentLeft = "true"
        android:layout_alignParentTop = "true"
        android:layout_marginLeft = "27dp"
        android:layout_marginTop = "31dp"
        android:text = "@string/txtUser" />
    <TextView
        android:id = "@+id/textView2"
        android:layout_width = "wrap_content"
        android:layout_height = "wrap_content"
        android:layout_alignLeft = "@+id/textView1"
        android:layout_below = "@+id/textView1"
        android:layout_marginTop = "38dp"
        android:text = "@string/txtPwd" />
    <EditText
        android:id = "@+id/etUserName"
        android:layout_width = "wrap_content"
        android:layout_height = "wrap_content"
        android:layout_alignBaseline = "@+id/textView1"
        android:layout_alignBottom = "@+id/textView1"
        android:layout_marginLeft = "40dp"
        android:layout_toRightOf = "@+id/textView1"
        android:ems = "10" >
        <requestFocus />
    </EditText>
    <EditText
        android:id = "@+id/etPwd"
        android:layout_width = "wrap_content"
        android:layout_height = "wrap_content"
        android:layout_alignBaseline = "@+id/textView2"
        android:layout_alignBottom = "@+id/textView2"
        android:layout_alignLeft = "@+id/etUserName"
        android:ems = "10"
        android:inputType = "textPassword" />
```

```
    < Button
        android:id = " @ + id/btnLogin"
        android:layout_width = " wrap_content"
        android:layout_height = " wrap_content"
        android:layout_alignLeft = " @ + id/etPwd"
        android:layout_below = " @ + id/etPwd"
        android:layout_marginTop = " 38dp"
        android:text = " @string/btnLogin" / >
    < Button
        android:id = " @ + id/btnReset"
        android:layout_width = " wrap_content"
        android:layout_height = " wrap_content"
        android:layout_alignBaseline = " @ + id/btnLogin"
        android:layout_alignBottom = " @ + id/btnLogin"
        android:layout_alignRight = " @ + id/etPwd"
        android:text = " @string/btnReset" / >
</RelativeLayout >
```

④在 CheckRadioActivity.java 中，修改代码，如下所示：

```
import android.app.Activity;
import android.os.Bundle;
import android.view.Menu;
import android.view.View;
import android.view.View.OnClickListener;
import android.widget.Button;
import android.widget.EditText;
import android.widget.Toast;
public class MainActivity extends Activity {
    EditText etUserName;
    EditText etPwd;
    Button btnLogin;
    Button btnReset;
    @Override
    protected void onCreate( Bundle savedInstanceState) {
        super.onCreate(savedInstanceState);
        setContentView( R.layout.activity_main);
        etUserName = ( EditText ) findViewById( R.id.etUserName);
        etPwd = ( EditText ) findViewById( R.id.etPwd);
        btnLogin = ( Button ) findViewById( R.id.btnLogin);
        btnReset = ( Button ) findViewById( R.id.btnReset);
        btnLogin.setOnClickListener( new OnClickListener() {
```

```java
            @Override
            public void onClick(View v) {
                // TODO Auto-generated method stub
                if(etUserName.getText().length() > 0
                    &&etPwd.getText().length() > 0){
                    Toast.makeText(MainActivity.this,"登录成功!",
                      Toast.LENGTH_LONG).show();
                }else{
                    Toast.makeText(MainActivity.this,
                    "用户名或密码不能为空!",Toast.LENGTH_LONG).show();
                }
            }
        });
        btnReset.setOnClickListener(new OnClickListener() {
            @Override
            public void onClick(View v) {
                // TODO Auto-generated method stub
                etUserName.setText("");
                etPwd.setText("");
            }
        });
    }
}
```

(2) 代码分析

例2-5中的程序首先通过 findViewById() 方法找到界面中的各个组件，然后，为 Button 设置监听器，处理用户单击事件。在事件处理程序中完成相应的程序逻辑。

(3) 功能扩展

修改例2-5的程序，当用户未输入信息时，在 EditText 中显示相关的提示信息，如图2.16所示。当用户登录成功时，显示"欢迎×××登录系统"的信息，其中"×××"为用户输入的用户名。

3) CheckBox 和 RadioButton

(1) CheckBox

复选框（CheckBox）继承自 CompoundButton，大部分属性与 Button 类似。它是一种有双状态按钮的特殊类型，可以选中或者不选中。可通过 isChecked() 方法判断是否选中。

(2) RadioButton

单选按钮（RadioButton）继承自 CompoundButton，大部分属性与 Button 类似。它是一种双状态的按钮，可以选中或不选中。在单选按钮没有被选中时，用户能够按下或单击来选中它。但是，与复选框相反，用户一旦选中就不能够取消选中（可以通过代码来控制，界面上点击的效果是一旦选中之后就不能取消选中了）。

多个单选按钮通常与 RadioGroup 同时使用。当一个单选按钮组（RadioGroup）包含几个单选按钮时，选中其中一个的同时将取消其他选中的单选按钮。可通过 isChecked( ) 方法判断是否选中。

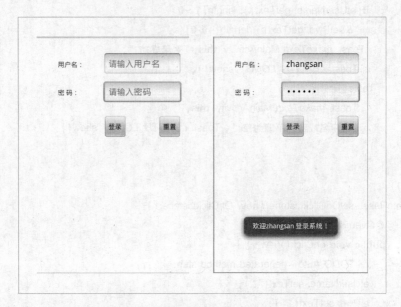

图 2.16 修改后的程序效果

【例 2-6】 实现图 2.17 所示的求职意向调查程序。用户选择性别和求职岗位信息后，单击"确定"按钮，在屏幕上方显示用户的选择信息。

（1）实现过程

①参照项目1，在 Eclipse 中创建名为 ex02_Jobs 的工程。

②打开 res\\values\\strings.xml 文件，写入如下内容：

<string name="*male*">男</string><string name="*female*">女</string>

<string name="*nochoice*">您还没有选择</string>

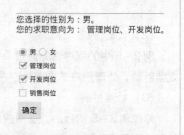

图 2.17 求职意向调查程序

<string name="*admin*">管理岗位</string>

<string name="*prog*">开发岗位</string>

<string name="*sales*">销售岗位</string>

<string name="*ok*">确定</string>

③修改布局文件 activity_main.xml，代码如下所示：

<?xml version="1.0" encoding="*utf-8*"?>

<LinearLayout xmlns:android="*http://schemas.android.com/apk/res/android*"

　　android:orientation="*vertical*"

　　android:layout_width="*fill_parent*"

```xml
        android:layout_height = "fill_parent" >
    <TextView
        android:text = "@string/nochoice"
        android:textSize = "20dip"
        android:id = "@+id/TextView01"
        android:layout_width = "wrap_content"
        android:layout_height = "wrap_content" >
    </TextView>
    <RadioGroup
        android:id = "@+id/RadioGroup01"
        android:orientation = "horizontal"
        android:layout_width = "wrap_content"
        android:layout_height = "wrap_content" >
        <RadioButton
            android:text = "@string/male"
            android:id = "@+id/male"
            android:checked = "true"
            android:layout_width = "wrap_content"
            android:layout_height = "wrap_content" >
        </RadioButton>
        <RadioButton
            android:text = "@string/female"
            android:id = "@+id/female"
            android:layout_width = "wrap_content" >
        </RadioButton>
    </RadioGroup>
    <LinearLayout
        android:id = "@+id/LinearLayout01"
        android:orientation = "vertical"
        android:layout_width = "wrap_content"
        android:layout_height = "wrap_content" >
        <CheckBox
            android:text = "@string/admin"
            android:id = "@+id/CheckBox01"
            android:layout_width = "wrap_content"
            android:layout_height = "wrap_content" >
        </CheckBox>
        <CheckBox
            android:text = "@string/prog"
            android:id = "@+id/CheckBox02"
```

```
                android:layout_width = "wrap_content"
                android:layout_height = "wrap_content" >
            </CheckBox>
            <CheckBox
                android:text = "@string/sales"
                android:id = "@+id/CheckBox03"
                android:layout_width = "wrap_content"
                android:layout_height = "wrap_content" >
            </CheckBox>
        </LinearLayout>
        <Button
            android:text = "@string/ok"
            android:id = "@+id/Button01"
            android:layout_width = "wrap_content"
            android:layout_height = "wrap_content" >
        </Button>
</LinearLayout>
```

④在 CheckRadioActivity.java 中，修改代码，如下所示：

```
import android.app.Activity;
import android.os.Bundle;
import android.view.View;
import android.view.View.OnClickListener;
import android.widget.Button;
import android.widget.CheckBox;
import android.widget.RadioButton;
import android.widget.TextView;
public class CheckRadioActivity extends Activity
{
    /** Called when the activity is first created. */
    RadioButton rb_male;              //男单选按钮
    CheckBox cb_admin;                //管理岗位 CheckBox
    CheckBox cb_prog;                 //开发岗位
    CheckBox cb_sales;                //销售岗位
    Button btn_ok;                    //确定按钮
    StringBuffer result;              //声明一个 StringBuffer
    TextView content;                 //TextView 中的显示文本
    @Override
    public void onCreate(Bundle savedInstanceState)
    {
        super.onCreate(savedInstanceState);
```

```java
setContentView(R.layout.main);        //设置布局文件
result = new StringBuffer();          //创建 StringBuffer 对象
//获取 TextView 的引用
content = (TextView)findViewById(R.id.TextView01);
//获取男单选按钮的引用
rb_male = (RadioButton)findViewById(R.id.male);
//确定按钮的引用
btn_ok = (Button)findViewById(R.id.Button01); btn_ok.setOnClickListener(
    //添加监听器
    new OnClickListener()
    {
        public void onClick(View v)
        {
            //获取三个求职意向 CheckBox 的引用
            cb_admin = (CheckBox)findViewById(R.id.CheckBox01);
            cb_prog = (CheckBox)findViewById(R.id.CheckBox02);
            cb_sales = (CheckBox)findViewById(R.id.CheckBox03);
            //创建一个字符串对象
            String result = "您选择的性别为:";
            if(rb_male.isChecked())
            {                                //若男单选按钮被选中,则性别为男
                result = result + "男。\n";
            }
            else                             //否则为女
            {
                result = result + "女。\\n";
            }
            String jobStr = "";              //创建求职意向字符串
            if(cb_admin.isChecked())         //设置字符串文本
            {
                jobStr += "管理岗位" + "、";
            }
            if(cb_prog.isChecked())
            {
                jobStr += "开发岗位" + "、";
            }
            if(cb_sales.isChecked())
            {
                jobStr += "销售岗位" + "、";
            }
```

```
            if( jobStr. length( ) > 0 )
            {       //设置 TextView 的显示文本
                    result = result + "您的求职意向为:
                    " + jobStr. substring( 0, jobStr. length( ) - 1 ) +
                    "。\\n";
            }
            else
            {
                    result = result + "。\\n";
            }
            //将处理好的字符串设置为 TextView 的显示内容
            content. setText( result) ;}
        }
    );
}
```

(2) 代码分析

例2-6 的程序基本结构与例2-5 相似,RadioButton 和 CheckBox 均可通过 isChecked( ) 方法获取是否为选中状态。

(3) 功能扩展

修改例2-6 的代码,当用户选择某个 CheckBox 时,利用 Toast 在屏幕上显示选择的是哪个选项。

提示:需要为每个 CheckBox 建立 setOnCheckedChangeListener 监听器。

4) ImageView

ImageView 用来显示任意图像图片,可以自己定义显示尺寸、显示颜色等。

(1) 主要属性

①android:adjustViewBounds 是否保持宽高比。需要与 maxWidth、MaxHeight 一起使用,单独使用没有效果。

②android:cropToPadding 是否截取指定区域用空白代替。单独设置无效,需要与 scrollY 一起使用。

③android:maxHeight 定义 View 的最大高度,需要与 AdjustViewBounds 一起使用,单独使用没有效果。如果想设置图片固定大小,又想保持图片宽高比,需要如下设置:

- 设置 AdjustViewBounds 为 true;
- 设置 maxWidth、MaxHeight;
- 设置 layout_width 和 layout_height 为 wrap_content。

④android:maxWidth 设置 View 的最大宽度。

⑤android:scaleType 设置图片的填充方式。

⑥android：src 设置 View 的图片或颜色。

⑦android：tint 将图片渲染成指定的颜色。

⑧android：scaleType：控制图片如何 resized/moved 来匹配 ImageView 的 size。ImageView.ScaleType / android：scaleType 值的意义如下：

- CENTER /center 按图片的原来 size 居中显示，当图片长/宽超过 View 的长/宽时，截取图片的居中部分显示。
- CENTER_CROP / centerCrop 按比例扩大图片的 size 居中显示，使得图片长（宽）等于或大于 View 的长（宽）。
- CENTER_INSIDE / centerInside 将图片的内容完整居中显示，通过按比例缩小原来的 size 使得图片长/宽等于或小于 View 的长/宽。
- FIT_CENTER / fitCenter 把图片按比例扩大/缩小到 View 的宽度，居中显示。
- FIT_END / fitEnd 把图片按比例扩大/缩小到 View 的宽度，显示在 View 的下部分位置。
- FIT_START / fitStart 把图片按比例扩大/缩小到 View 的宽度，显示在 View 的上部分位置。
- FIT_XY / fitXY 把图片不按比例扩大/缩小到 View 的大小显示。
- MATRIX / matrix 用矩阵来绘制、动态缩小/放大图片来显示。

（2）常用方法

①public void setVisibility（int visibility），其中 visibility 是 int 型的参数，取值分别为 VISIBLE、INVISIBLE 和 GONE。

对应上面：VISIBLE＝0x00000000；INVISIBLE＝0x00000004；GONE＝0x00000008。即：

image.setVisibility(0x00000000)

image.setVisibility(View.VISIBLE);   //表示显示

image.setVisibility(0x00000004)

image.setVisibility(View.INVISIBLE);   //表示隐藏

image.setVisibility(0x00000008)

image.setVisibility(View.GONE);   //表示 view 不存在

②设置颜色的不同方法。

color.rgb(255,255,255);

color.RED;

color.parseColor(colorString);

//其中 colorString 可以是#RRGGBB #AARRGGBB 'red', 'blue',

// 'green', 'black', 'white', 'gray', 'cyan', 'magenta',

// 'yellow', 'lightgray', 'darkgray' 等

③设置图片指定大小。

protected Bitmap scaleImg(Bitmap bm, int newWidth, int newHeight) {

　　// 图片源

　　Bitmap bm = BitmapFactory.decodeStream(getResources().openRawResource(id));

　　// 获得图片的宽高

```
        int width = bm.getWidth();
        int height = bm.getHeight();
        //设置想要的大小
        int newWidth1 = newWidth;
        int newHeight1 = newHeight;
        //计算缩放比例
        float scaleWidth = ((float)newWidth1) / width;
        float scaleHeight = ((float)newHeight1) / height;
        //取得想要缩放的matrix参数
        Matrix matrix = new Matrix();
        matrix.postScale(scaleWidth, scaleHeight);
        //得到新的图片
        Bitmap newbm = Bitmap.createBitmap(bm, 0, 0, width, height, matrix, true);  return newbm;
}
//调用
//获得18×18的图片
Bitmap bm = BitmapFactory.decodeStream(getResources().openRawResource(R.drawable.icon));
Bitmap newBm = scaleImg(bmImg, 18, 18);
imageView.setImageBitmap(newBm);
```

【例 2-7】 实现图 2.18 所示的图片浏览程序。

（1）实现过程

① 参照项目 1，在 Eclipse 中创建名为 ex02_showPhoto 的工程。

② 打开 res\\values\\strings.xml 文件，写入如下内容：

```
<string name="pre">上一幅</string>
<string name="after">下一幅</string>
```

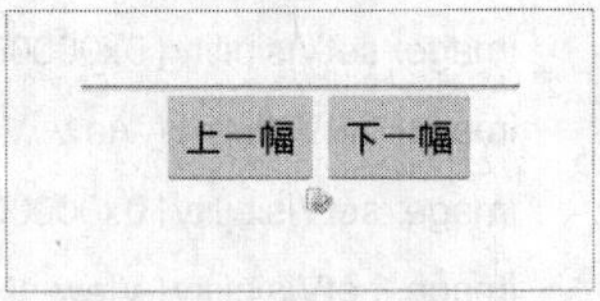

③ 将图片文件复制到 res\\drawable-mdpi 文件夹下。

图 2.18　图片浏览程序

④ 修改布局文件 activity_main.xml，代码如下所示：

```
<?xml version="1.0" encoding="utf-8"?>
<LinearLayout xmlns:android="http://schemas.android.com/apk/res/android"
    android:orientation="vertical"
    android:layout_width="fill_parent"
    android:layout_height="fill_parent">
    <LinearLayout
        android:id="@+id/LinearLayout01"
        android:orientation="horizontal"
        android:layout_width="wrap_content"
        android:layout_gravity="center"
```

```xml
        android:layout_height = "wrap_content" >
        <Button
            android:text = "@string/pre"
            android:id = "@+id/Button01"
            android:textColor = "#000000"
            android:layout_width = "wrap_content"
            android:layout_height = "wrap_content" >
        </Button>
        <Button
            android:text = "@string/after"
            android:id = "@+id/Button02"
            android:textColor = "#000000"
            android:layout_width = "wrap_content"
            android:layout_height = "wrap_content" >
        </Button>
    </LinearLayout>
    <ImageView android:id = "@+id/ImageView01"
        android:layout_width = "wrap_content"
        android:layout_height = "wrap_content"
        android:layout_gravity = "center" >
    </ImageView>
</LinearLayout>
```

⑤在ImageViewActivity.java中，修改代码，如下所示：

```java
import android.app.Activity;
import android.os.Bundle;
import android.view.View;
import android.view.View.OnClickListener;
import android.widget.Button;
import android.widget.ImageView;
import android.widget.Toast;
public class ImageViewActivity extends Activity {
    /** Called when the activity is first created. */
    ImageView iv;                    //声明ImageView控件
    int count;                       //计数器
    int drawableIds[ ] = {           //图片Id数组
        R.drawable.pic0,
        R.drawable.pic1,
        R.drawable.pic2
    };
```

```java
@Override
public void onCreate(Bundle savedInstanceState) {
    super.onCreate(savedInstanceState);
    setContentView(R.layout.activity_main);         //设置布局
    //获取ImageView的引用
    iv = (ImageView)findViewById(R.id.ImageView01);
    iv.setImageResource(R.drawable.pic0);           //设置图像
    //获取前一个按钮的引用
    Button pre_btn = (Button)findViewById(R.id.Button01);
    pre_btn.setOnClickListener(//添加监听器
        new OnClickListener()
        {
            @Override
            public void onClick(View v) {
                if(count > 0)
                {    //设置图像
                    iv.setImageResource(drawableIds[--count]);
                }
                else
                {
                    Toast.makeText(ImageViewActivity.this,
                    "这是第一幅图片",Toast.LENGTH_SHORT).show();
                }
            }
        }
    );
    //获取下一个按钮的引用
    Button after_btn = (Button)findViewById(R.id.Button02);
    after_btn.setOnClickListener(//添加监听器
    new OnClickListener()
    {
        @Override
        public void onClick(View v) {
            if(count < drawableIds.length-1)
            {    //设置图像
                iv.setImageResource(drawableIds[++count]);
            }
            else
            {
```

```
                    Toast. makeText( ImageViewActivity. this,
                        "这是最后一幅图片",Toast. LENGTH_SHORT). show( );
                    }
                }
            }
        );
    }
}
```

(2) 代码分析

本程序将图片的 id 放入数组中，通过更新指向数组当前位置的变量 count 的值，达到前后浏览图片的功能。

①当单击"上一幅"按钮时，iv. setImageResource（drawableIds［－－count］）；

②当单击"下一幅"按钮时，iv. setImageResource（drawableIds［＋＋count］）；

(3) 功能拓展

修改以上例子，添加图片放大功能，效果如图 2.19 所示。

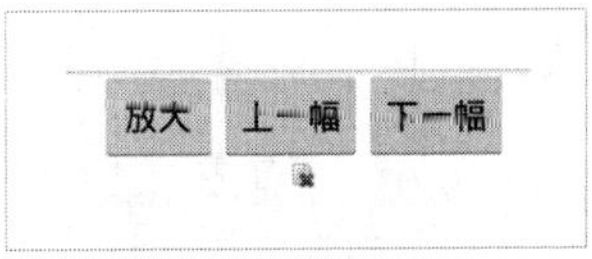

提示：

可以使用 Martix（android. graphics. Matrix） 类中的 postScale（ ）方法结合 Bitmap 来实现缩放图片的功能。关键代码如下：

图 2.19　带图片放大功能的浏览程序

Bitmap bmp ＝ BitmapFactory. decodeResource（getResource（），R. drawalbe. icon1）

int bmpwidth ＝ bmp. getWidth( )；

int bmpheight ＝ bmp. getHeight( )；

Matrix matrix ＝ new Matrix( )；

matrix. postScale(scaleW，scaleH)；

Bitmap bm ＝ Bitmap. createBitmap(bmp,0,0,bmpwidth,bmpheight ,matrix,true)；

imageView. setImageBitmap(bm)；

其中，scaleW 和 scaleH 为缩放系数，float 类型，1 表示原大小。注意：在 Android 中不允许 ImageView 在产生后动态修改其长度和宽度，所以要实现图片放大/缩小的功能，必须将原来的 ImageView 移除，重新产生一个新的 ImageView，并且为其指定图片来源，再将其放入 Layout 中。

任务 1　实现仿 QQ 登录基本界面

1. 任务说明

本项目实现了一个仿 QQ 登录的界面，效果如图 2.20 所示。

2. 实现过程

①参照项目 1，在 Eclipse 中创建名为 QQDemoV1 的工程，将默认创建的 MainActivity 重命

名为 LoginActivity。

②打开 res\\values\\strings.xml 文件，写入如下内容：

<？xml version ="1.0" encoding ="utf-8"？>
<resources>
    <string name ="app_name">QQDemoV1</string>
    <string name ="action_settings">Settings</string>
    <string name ="txt_qq">请输入 QQ 号</string>
    <string name ="txt_pwd">请输入密码</string>
    <string name ="btn_login">登录</string>
    <string name ="chk_remember">记住密码</string>
    <string name ="btn_register">注册新用户</string>
</resources>

图 2.20　仿 QQ 登录界面

③将图片资源复制到 res\\drawable-hdpi 中，如图 2.21 所示。

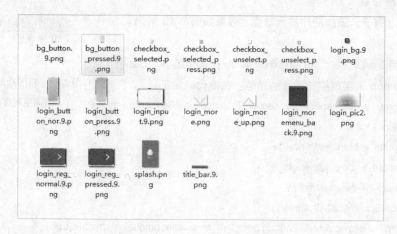

图 2.21　项目中需要使用的图片资源

④新建布局文件 activity_login.xml，设置为相对布局。设置 android：background ="@drawable/login_bg"，如图 2.22 所示。

⑤拖动 ImageView 组件放到界面中，分别设置以下属性，效果如图 2.23 所示。

android：src ="@drawable/login_pic2"
android：layout_alignParentTop ="true"
android：layout_centerHorizontal ="true"
android：layout_marginTop ="5dp"

⑥对于比较复杂的界面设计，需要借助 Outline 视图。选择 Windows→Show View→Outline 命令，显示 Outline 视图，可看到界面的结构图，将 LinearLayout 拖放到 ImageView 下方，如图 2.24 所示。

⑦依次拖放相应的组件到适当的位置，并设置对应的属性，布局

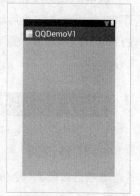

图 2.22　设置背景图片的界面效果

结构如图 2.25 所示，显示效果如图 2.26 所示。

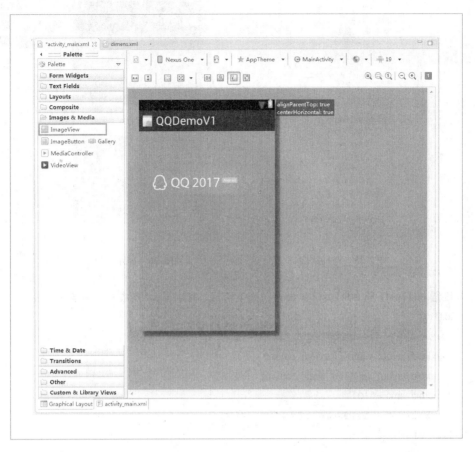

图 2.23 设置 ImageView 的属性

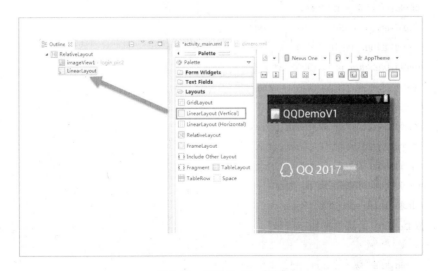

图 2.24 打开 Outline 视图

图 2.25 布局结构

图 2.26 显示效果

⑧设置 editText1 和 editText2 的属性，如下所示。其中 View 用来显示一条分隔线。

<EditText

  android:id = " @ + id/editText1 "

  android:layout_width = " match_parent"

  android:layout_height = " wrap_content"

  android:ems = " 10"

  android:hint = " @string/txt_qq"   >

</EditText >

<View

  android:id = " @ + id/view1 "

  android:layout_width = " wrap_content"

  android:layout_height = " 1 dp"

  android:background = " #333333"   / >

<EditText

  android:id = " @ + id/editText2"

  android:layout_width = " match_parent"

  android:layout_height = " wrap_content"

  android:ems = " 10"

  android:hint = " @string/txt_pwd"

  android:inputType = " textPassword"   / >

⑨设置 button1 的属性，如下所示。

<Button

  android:id = " @ + id/button1 "

  android:layout_width = " match_parent"

```
android:layout_height = "wrap_content"
android:layout_below = "@+id/linearLayout1"
android:layout_centerHorizontal = "true"
android:layout_marginTop = "10dp"
android:background = "@drawable/login_button_nor"
android:text = "@string/btn_login" />
```

⑩设置 checkBox1 的属性，如下所示。

```
<CheckBox
    android:id = "@+id/checkBox1"
    android:layout_width = "wrap_content"
    android:layout_height = "wrap_content"
    android:layout_alignLeft = "@+id/button1"
    android:layout_below = "@+id/button1"
    android:layout_marginTop = "10dp"
    android:button = "@null"
    android:drawableLeft = "@drawable/checkbox_unselect"
    android:text = "@string/chk_remember"
    android:textColor = "#ffffff"
    android:textSize = "14sp" />
```

⑪设置 button2 的属性，如下所示。

```
<Button
    android:id = "@+id/button2"
    android:layout_width = "110dp"
    android:layout_height = "33dp"
    android:layout_alignRight = "@+id/button1"
    android:layout_alignTop = "@+id/checkBox1"
    android:background = "@drawable/login_reg_normal"
    android:paddingRight = "10dp"
    android:text = "@string/btn_register"
    android:textColor = "#ffffff"
    android:textSize = "14sp"
/>
```

⑫界面效果如图 2.27 所示。

⑬在模拟器中运行，发现 checkBox 无法选中，因为 android:drawableLeft = "@drawable/checkbox_unselect"，这意味着 checkBox 永远只显示这张图片，因此无法显示选中状态，同时界面中的按钮也没有点击的动态效果。为解决这些问题，需要使用 selector（即选择器）。在 Android 中常常用 Selector 来作组件的背景，这样做的好处是省去了用代码控制实现组件在不同状态下不

图 2.27 界面效果

同的背景颜色或图片的变换,使用十分方便。右击 res\\drawable 目录,在快捷菜单中选择 New→Android XML File 命令,系统弹出对话框,选择 selector,输入名字 chk_remember_selector,如图 2.28 所示。

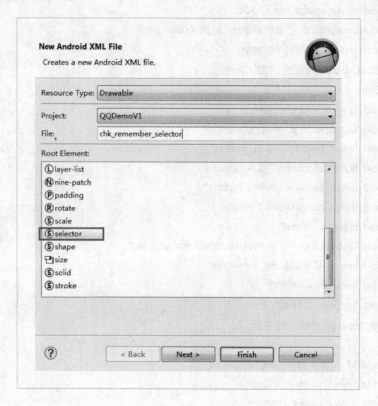

图 2.28　新建 selector

⑭在该文件中输入如下内容。
< ? xml version = "1.0" encoding = "utf-8" ? >
< selector xmlns:android = "http://schemas.android.com/apk/res/android" >
　　< item android:state_checked = "true" android:state_selected = "true" android:drawable = "@drawable/checkbox_selected_press" > </item >
　　< item android:state_checked = "true" android:state_selected = "false" android:drawable = "@drawable/checkbox_selected" > </item >
　　< item android:state_checked = "false" android:state_selected = "true" android:drawable = "@drawable/checkbox_unselect_press" > </item >
　　< item android:state_checked = "false" android:state_selected = "false" android:drawable = "@drawable/checkbox_unselect" > </item >
</selector >
⑮修改 checkBox1 的属性。
android:drawableLeft = "@drawable/chk_remember_selector"

⑯参照第⑬和⑭步，新建 btn_login_selector.xml。内容如下：

<?xml version="1.0" encoding="utf-8"?>
<selector xmlns:android="http://schemas.android.com/apk/res/android">
    <item android:state_pressed="true"   android:drawable="@drawable/bg_button_pressed">
    </item>
    <item android:state_pressed="false" android:drawable="@drawable/login_button_nor">
    </item>
</selector>

⑰参照第⑬和⑭步，新建 btn_register_selector.xml。内容如下：

<?xml version="1.0" encoding="utf-8"?>
<selector xmlns:android="http://schemas.android.com/apk/res/android">
    <item android:state_pressed="true"   android:drawable="@drawable/login_reg_pressed">
    </item>
    <item android:state_pressed="false" android:drawable="@drawable/login_reg_normal">
    </item>
</selector>

⑱分别修改 button1 和 button2 中的 android：background 属性值。

android:background="@drawable/btn_login_selector"

android:background="@drawable/btn_register_selector"

⑲最终显示效果如图 2.29 所示。

图 2.29　最终显示效果

3. 代码分析

①Android 布局使用 layout 中的 XML 文件实现，排版逻辑大体类似于 div+css 结构，对于

较为复杂的界面,可以通过多种布局嵌套实现。在本任务中,通过在相对布局中嵌套线性布局实现不同的背景色块。具体结构如图 2.25 所示。

②选择器 selector.xml。通过设置 selector.xml 可使得控件在不同操作下(默认、点击、焦点等)显示不同样式,基本属性如表 2.1 所示。其中,有关状态属性的取值皆为 boolean 属性:true、false。

表 2.1　selector 属性

XML 属性	说　　明
android:drawable	放一个 drawable 资源
android:state_pressed	按下状态,如一个按钮触摸或者点击
android:state_focused	取得焦点状态,比如用户选择了一个文本框
android:state_hovered	光标悬停状态,通常与 focused state 相同,它是 4.0 的新特性
android:state_selected	选中状态
android:state_enabled	能够接收触摸或者点击事件
android:state_checked	被 checked 了,如一个 RadioButton 可以被 check

在 drawable 中添加 selector.xml 资源文件,下面的代码使按钮按下和松开显示不同的图片。

<?xml version = "1.0" encoding = "UTF-8"?>

<selector xmlns:android = "http://schemas.android.com/apk/res/android">

<!-- 指定按钮按下时的图片 -->

<item android:state_pressed = "true"　android:drawable = "@drawable/start_down"/>

<!-- 指定按钮松开时的图片 -->

<item android:state_pressed = "false"　android:drawable = "@drawable/start"/>

</selector>

任务 2　实现界面的动态展示

1. 任务说明

本任务在任务 1 的基础上,增加了可伸缩的选项面板,效果如图 2.30 所示。

2. 实现过程

①在 QQDemoV1 的工程中,打开布局文件 activity_login.xml,添加垂直方向线性布局 LinearLayout,设置 android:layout_alignParentBottom = "true"。

<LinearLayout

　　android:layout_width = "fill_parent"

　　android:layout_height = "wrap_content"

　　android:layout_alignParentBottom = "true"

　　android:layout_centerHorizontal = "true"

android：background = " @drawable/login_moremenu_back"

android：orientation = " vertical" >

</LinearLayout >

图 2.30　可伸缩面板效果

②在该布局中添加一个相对布局（menu_more）和一个线性布局 menu_panel，如图 2.31 所示。

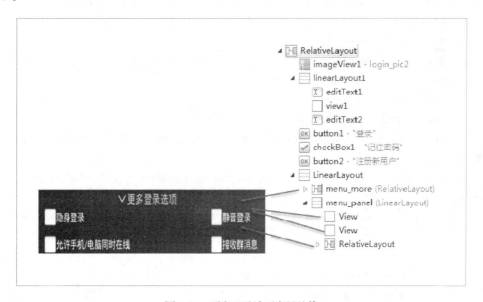

图 2.31　更多登录选项布局结构

③完成相对布局 menu_more 的内部设计。注意，要使该布局可被点击，需设置 android：clickable = " true"。具体代码如下：

```xml
<RelativeLayout android:id="@+id/menu_more"
    android:layout_width="fill_parent"
    android:clickable="true"
    android:layout_height="20dp" >
    <TextView    android:id="@+id/textView1"
        android:layout_width="wrap_content"
        android:layout_height="wrap_content"
        android:layout_alignParentBottom="true"
        android:layout_centerHorizontal="true"
        android:text="@string/txt_menumore"
        android:background="@null"
        android:textColor="#c6e6f9"
        android:textSize="14sp" />
    <ImageView    android:id="@+id/imageView2"
        android:layout_width="wrap_content"
        android:layout_height="wrap_content"
        android:layout_centerVertical="true"
        android:layout_toLeftOf="@+id/textView1"
        android:src="@drawable/login_more_up" />
</RelativeLayout>
```

④完成线性布局 menu_panel 的内部设计，代码如下。注意，此处通过设置 visibility 属性控制选项面板是否展开。开始时，设置 android：visibility="gone"，使选项面板隐藏。

```xml
<LinearLayout
    android:id="@+id/menu_panel"
    android:layout_width="fill_parent"
    android:layout_height="wrap_content"
    android:visibility="gone"
    android:orientation="vertical"   >
    <View android:layout_width="fill_parent" android:layout_height="1.0px"    android:background="#005484" />
    <View android:layout_width="fill_parent" android:layout_height="1.0px"    android:background="#0883cb" />
    <RelativeLayout android:layout_width="fill_parent"    android:layout_height="wrap_content" >
        <CheckBox
            android:id="@+id/checkBox3"
            android:layout_width="wrap_content"
            android:layout_height="wrap_content"
            android:layout_alignParentLeft="true"
            android:button="@null"
            android:drawableLeft="@drawable/chk_remember_selector"
```

```xml
            android:textColor = "#ffffff"
            android:textSize = "12sp"
            android:layout_marginLeft = "10dp"
            android:text = "@string/chk_hidelogin" />
        <CheckBox
            android:id = "@+id/checkBox4"
            android:layout_width = "wrap_content"
            android:layout_height = "wrap_content"
            android:layout_alignParentRight = "true"
            android:button = "@null"
            android:drawableLeft = "@drawable/chk_remember_selector"
            android:layout_marginRight = "25dp"
            android:text = "@string/chk_silentlogin"
            android:textColor = "#ffffff"
            android:textSize = "12sp" />
        <CheckBox
            android:id = "@+id/checkBox5"
            android:layout_width = "wrap_content"
            android:layout_height = "wrap_content"
            android:layout_alignLeft = "@+id/checkBox3"
            android:layout_below = "@+id/checkBox3"
            android:button = "@null"
            android:drawableLeft = "@drawable/chk_remember_selector"
            android:paddingTop = "10dp"
            android:text = "@string/chk_bothonline"
            android:textColor = "#ffffff"
            android:textSize = "12sp" />
        <CheckBox
            android:id = "@+id/checkBox6"
            android:layout_width = "wrap_content"
            android:layout_height = "wrap_content"
            android:layout_alignLeft = "@+id/checkBox4"
            android:layout_below = "@+id/checkBox4"
            android:button = "@null"
            android:drawableLeft = "@drawable/chk_remember_selector"
            android:paddingTop = "10dp"
            android:text = "@string/chk_receivegroupmsg"
            android:textColor = "#ffffff"
            android:textSize = "12sp" />
    </RelativeLayout>
</LinearLayout>
```

⑤切换到 LoginActivity.java，在 MainActivity 类中添加成员变量 menu_more、img 和 menu_panel 分别对应布局文件中的相对布局 menu_more、menu_more 中的 imageView2 和线性布局 menu_panel。

    private LinearLayout menu_panel;
    private RelativeLayout menu_more;
    private ImageView img;

⑥修改 onCreate（Bundle savedInstanceState）方法中的代码，如下所示。

```
protected void onCreate(Bundle savedInstanceState) {
    super.onCreate(savedInstanceState);
    setContentView(R.layout.activity_main);
    img = (ImageView) findViewById(R.id.imageView2);
    menu_panel = (LinearLayout) findViewById(R.id.menu_panel);
    menu_more = (RelativeLayout) findViewById(R.id.menu_more);
    menu_more.setOnClickListener(new OnClickListener() {
        @Override
        public void onClick(View arg0) {
            // TODO Auto-generated method stub
            if(menu_panel.getVisibility() == View.GONE){
                menu_panel.setVisibility(View.VISIBLE);
                img.setImageResource(R.drawable.login_more);
            }else{
                menu_panel.setVisibility(View.GONE);
                img.setImageResource(R.drawable.login_more_up);
            }
        }
    });
}
```

⑦运行效果如图 2.30 所示。

⑧观察界面中的 CheckBox 组件，有多个属性值是相同的，例如：

android:layout_width = "*wrap_content*"
android:layout_height = "*wrap_content*"
android:button = "@*null*"
android:drawableLeft = "@*drawable/chk_remember_selector*"
android:textColor = "#*ffffff*"
android:textSize = "12*sp*"

⑨可以用将这些重复的属性的定义写在 style 文件中，来避免重复操作。打开 res\values\目录下的 styles.xml 文件，如下所示。

<resources xmlns:android = "*http://schemas.android.com/apk/res/android*">
    <!--

```xml
    Base application theme, dependent on API level. This theme is replaced
    by AppBaseTheme from res/values-vXX/styles.xml on newer devices.
-->
<style name="AppBaseTheme" parent="android:Theme.Light">
    <!--
        Theme customizations available in newer API levels can go in
        res/values-vXX/styles.xml, while customizations related to
        backward-compatibility can go here.
    -->
</style>
<!-- Application theme. -->
<style name="AppTheme" parent="AppBaseTheme">
    <!-- All customizations that are NOT specific to a particular API-level can go here. -->
</style>
</resources>
```

⑩在<resources></resources>节中添加自定义的style，如下所示。

```xml
<style name="chk_style" parent="@android:style/Widget.CompoundButton.CheckBox">
    <item name="android:layout_width">wrap_content</item>
    <item name="android:layout_height">wrap_content</item>
    <item name="android:drawableLeft">@drawable/chk_remember_selector</item>
    <item name="android:textColor">#fff</item>
    <item name="android:textSize">14sp</item>
    <item name="android:button">@null</item>
</style>
```

⑪修改布局文件activity_login.xml中CheckBox的属性，如下所示。

- 使用style之前：

```xml
<CheckBox
    android:id="@+id/checkBox1"
    android:layout_width="wrap_content"
    android:layout_height="wrap_content"
    android:layout_alignLeft="@+id/button1"
    android:layout_below="@+id/button1"
    android:layout_marginTop="10dp"
    android:button="@null"
    android:drawableLeft="@drawable/chk_remember_selector"
    android:text="@string/chk_remember"
    android:textColor="#ffffff"
    android:textSize="14sp" />
```

• 使用 style 之后：
```
<CheckBox
    android:id="@+id/checkBox1"
    style="@style/chk_style"
    android:layout_alignLeft="@+id/button1"
    android:layout_below="@+id/button1"
    android:layout_marginTop="10dp"
    android:text="@string/chk_remember"
/>
```

3. 代码分析

①Android 的样式一般定义在 res/values/styles.xml 文件中，其中有一个根元素 <resource>，而具体的每种样式定义则是通过 <resource> 下的子标签 <style> 来完成，<style> 通过添加多个 <item> 来设置样式不同的属性。

②样式是可以继承的，可通过 <style> 标签的 parent 属性声明要继承的样式，也可通过点前缀（.）继承，点前面为父样式名称，后面为子样式名称。点前缀方式只适用于自定义的样式，若要继承 Android 内置的样式，则只能通过 parent 属性声明。例如，在前面定义的 style 中，指定

parent="@android:style/Widget.CompoundButton.CheckBox"

③引用样式时只要在相应的组件里添加 style 即可，如在 CheckBox 引用 chk_style。

style="@style/chk_style"

④Activity 在 onCreate()方法中，通过 setContentView 实例化 layout 描述的控件。之后，通过 findViewById 获取控件实例。例如，要获取 imageView2，首先需要定义成员变量 img。

private ImageView img;

然后在 onCreate()方法中，通过 findViewById 获取该控件实例。

img=(ImageView)findViewById(R.id.imageView2);

任务3　实现欢迎界面

1. 任务说明

本任务在任务2的基础上，添加欢迎界面，效果如图2.32所示。

2. 实现过程

①新建布局文件 activity_welcome.xml，设置背景为 "@drawable/splash"，代码如下：

```
<?xml version="1.0" encoding="utf-8"?>
<FrameLayout xmlns:android="http://schemas.android.com/apk/res/android"
    android:layout_width="match_parent"
    android:layout_height="match_parent"
    android:background="@drawable/splash"
```

>

</FrameLayout>

②新建 WelcomeActivity 类继承自 Activity，重写 onCreate( )方法，代码如下：

public class WelcomeActivity extends Activity {
　　@Override
　　protected void onCreate(Bundle savedInstanceState) {
　　　　// TODO Auto-generated method stub
　　　　super.onCreate(savedInstanceState);
　　　　setContentView(R.layout.activity_welcome);
　　}
}

图 2.32　欢迎界面效果

③在 WelcomeActivity 类中添加 goLoginActivity( ) 方法，跳转到 LoginActivity，并关闭 welcomeActivity。

public void goLoginActivity( )
{
　　Intent intent = new Intent( );
　　intent.setClass(this, LoginActivity.class);
　　startActivity(intent);
　　finish( );
}

④在 WelcomeActivity 类中添加成员变量 mHandler 和 initView( ) 方法，实现显示 WelcomActiviy 界面 1 s 后，跳转到 LoginActivity 界面功能，代码如下：

```java
private Handler mHandler;
    public void initView() {
    
      mHandler = new Handler();
      mHandler.postDelayed(new Runnable() {
          @Override
          public void run() {
              // TODO Auto-generated method stub
              goLoginActivity();
          }
      }, 1000);
    }
```

⑤在onCreate()方法中,添加对方法initView()的调用,如下所示。

```java
public void onCreate(Bundle savedInstanceState) {
    super.onCreate(savedInstanceState);
    setContentView(R.layout.main);
    initView();
}
```

⑥打开AndroidManifest.xml文件,声明WelcomeActivity和LoginActivity,并指定WelcomeActivity为启动Activity,代码修改如下:

```xml
<application
        android:allowBackup="true"
        android:icon="@drawable/ic_launcher"
        android:label="@string/app_name"
        android:theme="@style/AppTheme" >
    <activity
            android:name="com.example.qqdemov1.WelcomeActivity"
            android:label="@string/app_name" >
        <intent-filter>
            <action android:name="android.intent.action.MAIN" />
            <category android:name="android.intent.category.LAUNCHER" />
        </intent-filter>
    </activity>
    <activity android:name="com.example.qqdemov1.LoginActivity"
            android:label="@string/app_name" >
    </activity>
</application>
```

3. 代码分析

①窗体跳转。Intent在Android开发中称为"意图",从字面意思不难理解,就是"你打算

去哪"。Intent 中有一个"栈"形式的容器,这个栈中存储的内容是一个个的 Activity,Activity 的 Start 和 Finish 操作对应着栈的 Push 和 Pop 操作。其常见写法如下:

Intent intent = new   Intent( );
intent. setClass(本类,将要跳转的类);
startActivity(intent);

②通过 Handler 对象调度线程。Handler 直接继承自 Object,一个 Handler 允许发送和处理 Message 或者 Runnable 对象,并且会关联到主线程的 MessageQueue 中。每个 Handler 具有一个单独的线程,并且关联到一个消息队列的线程,即 Handler 有一个固有的消息队列。当实例化一个 Handler 的时候,它就承载一个消息队列的线程,这个 Handler 可以把 Message 或 Runnable 压入消息队列,并且从消息队列中取出 Message 或 Runnable,进而操作它们。在本程序中,通过 postDelayed(Runnable r,long delayMillis)方法,把一个 Runnable 入队到消息队列中,UI 线程从消息队列中取出这个对象后,延迟 delayMills 秒执行。从而实现延时跳转功能。

实战演练

实现 QQ 注册界面,具体要求如下:
①实现 QQ 注册界面。
②在登录界面中实现点击"注册新用户"按钮,跳转到 QQ 注册界面的功能,如图 2.33 所示。

图 2.33 实战演练效果

项目 3 Android 高级 UI 组件——应用商店

 项目要点

- 了解适配器控件。
- 熟练使用适配器组件。
- 自定义适配器组件。

前面学习了基本 UI 组件、菜单组件，那些组件只能显示单一信息，有些情况下，需要显示同类多条信息，并以不同形式组织显示，需要高级 UI 组件中的适配器组件。本项目通过应用商店案例讲解高级 UI 适配器组件的使用，主要包括下拉列表控件 Spinner、列表视图 ListView、网格视图 GridView、画廊控件 Gallery 和选项卡 TabHost。

 项目简介

如图 3.1 所示，该项目综合运用高级图形 UI 控件，搭建一个完整的应用商店 UI 界面。

图 3.1 应用商店 UI 界面

 相关知识

1. 认识适配器控件

如图 3.2 所示，AdapterView 以合适的方式显示并操作一些数据（数组、链表、数据库等）。

适配器控件是一组可以通过特定的适配器将适配器控件的子控件与特定数据绑定起来的控件。适配器控件遵循 MVC 思想，其中适配器控件类似于视图，主要是呈现的框架（以下拉、列表、网格或者滚动画廊方式实现）。适配器控件就是控制器，主要控制框架中多个组件的显示内容和显示样式，其中 Model 以集合类数据对象的方式存在。

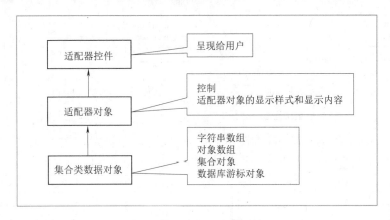

图 3.2　适配器控件

1）认识 Android 的适配器

Adapter 是 Android 中的重要角色，它是数据和 UI（View）之间的一个重要纽带。如图 3.3 所示，Adapter 负责创建用来表示每一个条目的 View 组件，并提供对底层数据的访问。支持 Adapter 绑定的控件必须覆盖 AdapterView 抽象类进行扩展。也可以创建由 AdapterView 派生的控件，并创建新的 Adapter 类来绑定它们。

一般地，不需要重新创建自己的 Adapter，Android 提供一个特定的 Adapter 向适配器控件提供数据。如上分析，Adapter 主要控制适配器控件上显示的数据和显示方式，所以可以定制它所绑定控件的外观和功能。其中每个条目可以相同，也可以不同。可以通过自定义 Adapter 定制每个条目。

一般情况下，使用特定的 Adapter 完成定制功能。

①ArrayAdapter 主要用于将集合对象中的每个对象的 toString 值，产生不同的 TextView 的对象，显示在适配器控件中。

②SimpleAdapter 主要用于在适配器控件中显示复杂的 View 对象，将集合对象中单个对象中的不同数据项填充到 View 中的不同组件中，并显示在适配器控件中的框架中。

③SimpleCursorAdapter 主要用于在适配器控件中显示复杂的 View 对象，通过将内容提供器返回的游标对象与 View 对象进行绑定，将游标对象中的不同数据项填充到 View 中的不同组件中，并显示在适配器控件中的框架中。

2）认识 Android 的适配器 UI 组件

如图 3.4 所示，适配器 UI 组件主要由适配器提供数据，显示和控制数据的框架，主要包括下拉列表控件 Spinner、列表视图 ListView、网格视图 GridView、画廊空间 Gallery 和选项卡 TabHost，其中 Spinner、ListView、GridView、Gallery 显示的基本原理相同，只是显示的框架不同，事件处理上会有所不同。

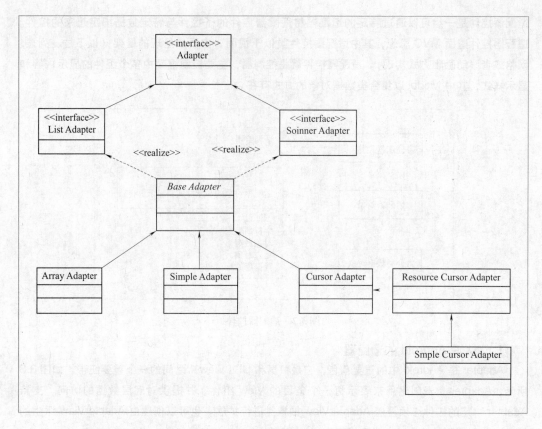

图 3.3 适配器类

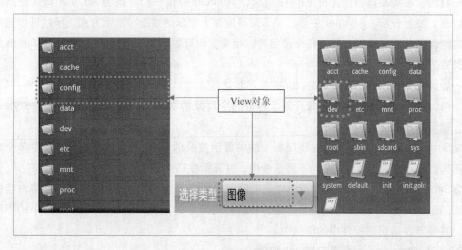

图 3.4 适配器控件框架

图 3.4 中分别是 Spinner、ListView 和 GridView 的显示样例，可以看到三种控件只是显示框架不同，每个框架中每个单元中放的还是 View 对象。View 对象的产生可以通过指定的 Adapter 对象自动产生，也可以通过自定义 Adapter 自定义 View 对象。

2. 下拉组件 Spinner

Spinner 是下拉列表,可以将多个 View 组件以下拉的形式组织起来,通常只显示选中的一项,当按下时才显示所有的选项。Spinner 是一个每次只能选择所有项中一项的组件。它的数据来源于与之关联的适配器,对下拉事件和下拉点击事件添加监听,实现对不同情况的不同处理。

1) Spinner 控件概述

Spinner 控件的类图如图 3.5 所示,从类图可以看出所以控件的父类都是 View 类,其代表矩形区域以及相应的事件处理。对于容纳其他 View 对象的容器类,View 有个子类 ViewGroup,该类是所有容器类控件的父类,这种容器包括各种布局类。本项目学到的 AdapterView 都是 ViewGroup 的子类。

```
java.lang.Object
  ↳ android.view.View
    ↳ android.view.ViewGroup
      ↳ android.widget.AdapterView<T extends android.widget.Adapter>
        ↳ android.widget.AbsSpinner
          ↳ android.widget.Spinner
```

图 3.5 Spinner 类图

Android 中的界面布局一般在 XML 文件中进行设置,所以所有控件都可以在 XML 文件中设置相应的属性,同时这些属性都对应类的 set( ) 方法。

①Spinner 常用属性和方法。

- android:dropDownHorizontalOffset

  对应方法:setDropDownHorizontalOffset(int)

  用于设置在 SpinnerMode 中 dropdown 下拉列表距离 spinner 控件的水平距离

- android:dropDownVerticalOffset

  对应方法:setDropDownVerticalOffset(int)

  用于设置在 SpinnerMode 中 dropdown 下拉列表距离 spinner 控件的垂直距离

- android:dropDownWidth

  对应方法:setDropDownWidth(int)

  用于设置 Spinner 的下拉列表的宽度。

- android:gravity

  对应方法:setGravity(int)

  用于定位当前选中项的 View 对象的相对位置。

- android:popupBackground

  对应方法:setPopupBackgroundResource(int)

  用于设置 Spinner 的下拉列表的背景图片。

- android:prompt

  Spinner 对话框显示的时候显示的提示信息。

- android:spinnerMode

②Spinner 可选项的显示模式。

对于 Spinner 中的一些方法，由于篇幅所限，大家自行查阅 API 文档，http：//developer. android. com/reference/android/widget/Spinner. html。

2）Spinner 事件处理

下拉选中事件：用于设置字体的颜色，如" #ff8c00"；

public void setOnItemSelectedListener（AdapterView. OnItemSelectedListener listener）

注册一个监听器，用于监听 Spinner 中的选项选中事件。

3）BaseAdapter

BaseAdapter 是一个公共基类适配器，实现了 SpinnerAdapter 和 ListAdapter，包含相应的方法，用于对 ListView 和 Spinner 等控件提供显示数据。BaseAdapter 的直接子类包括 ArrayAdapter、CursorAdapter 和 SimpleAdapter。

方法列表中，getCount（）方法用于返回适配器提供的 View 组件的数量，getView（）方法用于返回适配器控件的 position 位置的 View 对象，以便适配器对适配器控件进行填充。

4）ArrayAdapter

ArrayAdapter 主要用于将集合对象中的每个对象的 toString 值，产生不同的 TextView 的对象，显示在适配器控件中。

（1）构造器方法

- ArrayAdapter（Context context, int resource）
- ArrayAdapter（Context context, int resource, int textViewResourceId）
- ArrayAdapter（Context context, int resource, T[ ] objects）
- ArrayAdapter（Context context, int resource, int textViewResourceId, T[ ] objects）
- ArrayAdapter（Context context, int resource, List < T > objects）
- ArrayAdapter（Context context, int resource, int textViewResourceId, List < T > objects）

以上都是 ArrayAdapter 的重载的构造器，主要以不同形式提供 Model 数据，显示样式。可参考：http://developer. android. com/reference/android/widget/ArrayAdapter. html#ArrayAdapter （android. content. Context，int）

（2）常用方法

- void　add（T object）在 Model 中添加新对象。
- void　addAll（Collection < ? extends T > collection）在 Model 数组的后面添加集合对象。
- void　addAll（T...items）将集合对象中的数据添加到 Model 中。
- void　clear（）清除列表中所有对象。
- static ArrayAdapter < CharSequence > createFromResource（Context context, int textArrayResId, int textViewResId）从外部 XML 文件中产生一个新的 ArrayAdapter，其中 textArrayResId 提供 Model 数据，text ViewResId 提供 View 对象样式。
- Context　getContext（）返回该数组适配器对应的上下文。
- int　getCount（）返回 Model 列表的数量。

- View getDropDownView（int position，View convertView，ViewGroup parent）返回数据集下拉列表中特定位置的 View 对象。
- T getItem（int position）返回特定位置的 View 对象。
- long getItemId（int position）返回特定位置 View 对象的绑定 ID 值。
- int getPosition（T item）返回 Model 数组中特定对象的位置。
- View getView（int position，View convertView，ViewGroup parent）返回特定位置的 View 对象。
- void insert（T object, int index）在 Model 的特定索引位置插入对象。
- void notifyDataSetChanged( )提示 Model 数据已经发生变化，需刷新控制器组件。
- void remove（T object）移除数组中的特定对象。
- void setDropDownViewResource（int resource）设置下拉列表中单项显示的样式的资源文件。
- void setNotifyOnChange（boolean notifyOnChange）设置在 add（T）、insert（T, int）、remove（T）、clear( )调用后，是否自动调用数据更新方法。
- void sort（Comparator <？ super T > comparator）使用特定的排序器对 Model 数据进行排序。

5）基本 Spinner 应用

【例 3 – 1】 实现图 3.6 所示的界面。Spinner 显示下拉信息，选中后显示所选项。

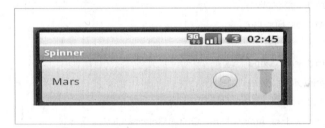

图 3.6　Ex03_01 程序

参照项目 1，在 Eclipse 中创建名为 Ex03_01 的工程。

（1）实现过程

①第一步。

编辑 res→layout 下的 activity_main. xml，增加两个 Spinner 控件，可以设置两个 Spinner 控件的 android：

layout_width = " match_parent"

android：layout_height = " wrap_content"

②第二步。

- 在 res→layout 下新建 spinnerlayout. xml，包含一个 TextView。

<？ xml version = " 1. 0" encoding = " utf-8" ？ >

<TextView

　　xmlns：android = " http：//schemas. android. com/apk/res/android"

```
        android:orientation = "vertical"
        android:layout_width = "wrap_content"
        android:layout_height = "wrap_content" android:id = "@ + id/textview1" >
</TextView>
```

- 在 res-values 下新建 array.xml，文件中定义 model 数据。

```
<? xml version = "1.0" encoding = "utf-8"? >
<resources>
    <string-array name = "countries">
        <item>北京</item>
        <item>上海</item>
        <item>广州</item>
        <item>深圳</item>
        <item>成都</item>
        <item>重庆</item>
    </string-array>
</resources>
```

③第三步。

编辑源文件 MainActivity.java。

1. **package** szpt.android.ex03_01；
2. **import** android.app.Activity；
3. **import** android.os.Bundle；
4. **import** android.util.Log；
5. **import** android.view.Menu；
6. **import** android.view.View；
7. **import** android.widget.AdapterView；
8. **import** android.widget.AdapterView.OnItemSelectedListener；
9. **import** android.widget.ArrayAdapter；
10. **import** android.widget.Spinner；
11.
12. **public class** MainActivity **extends** Activity {
13.   **private static final** String[] *Countries* = {"中国","美国","英国","日本","韩国",
14.   "瑞士"};
15.   Spinner spinner1 = **null**；
16.   Spinner spinner2 = **null**；
17.   ArrayAdapter<String> adapter = **null**；
18.
19.   @Override
20.   **protected void** onCreate(Bundle savedInstanceState) {
21.   **super**.onCreate(savedInstanceState)；
22.   setContentView(R.layout.*activity_main*)；
```
```

```
23.        spinner1 = (Spinner) findViewById(R.id.spinner1);
24.        adapter = new ArrayAdapter<String>(this,
25.                android.R.layout.simple_spinner_item, Countries);
26.        adapter.setDropDownViewResource(android.R.layout.simple_spinner_dropdown_item);
27.        spinner1.setAdapter(adapter);
28.        spinner1.setOnItemSelectedListener(new OnItemSelectedListener() {
29.            public void onItemSelected(AdapterView<?> arg0, View arg1,
30.                    int arg2, long arg3) {
31.                // TODO Auto-generated method stub
32.                Log.v("取出的数据是:", Countries[arg2]);
33.            }
34.
35.            public void onNothingSelected(AdapterView<?> arg0) {
36.                // TODO Auto-generated method stub
37.            }
38.        });
39.        spinner2 = (Spinner) findViewById(R.id.spinner2);
40.        ArrayAdapter<CharSequence> adapter2 = ArrayAdapter.createFromResource(
41.                this, R.array.countries, android.R.layout.simple_spinner_item);
42.        adapter2.setDropDownViewResource(R.layout.spinner_layout);
43.        spinner2.setAdapter(adapter);
44.    }
45.    @Override
46.    public boolean onCreateOptionsMenu(Menu menu) {
47.        // Inflate the menu; this adds items to the action bar if it is present.
48.        getMenuInflater().inflate(R.menu.main, menu);
49.        return true;
50.    }
51. }
```

(2) 代码分析

①第一步 activity_main.xml 中增加了两个 Spinner 控件，id 分别为 spinner1 和 spinner2。

②第二步中 spinnerlayout.xml 用来自定义 Spinner 下拉以后单个 View 的显示样式，array.xml 用 XML 文件定义 Spinner 的数据源。

③第三步第 13~17 行，Countries 定义了 spinner1 的数据源。

④第三步第 22~38 行取出布局文件中的 spinner1 存入成员变量中；

adapter = newArrayAdapter<String>(this, android.R.layout.simple_spinner_item, Countries);

⑤用于产生数组适配器，其中第二个参数使用系统自带的布局文件，指定 Spinner 中 View 对象的显示样式，第三个参数是适配器控件的 Model。

- adapter. setDropDownViewResource（android. R. layout. *simple_spinner_dropdown_item*）；使用系统自带的布局文件设置 *Spinner* 下拉后每个条目的显示样式。
- spinner1. setAdapter（adapter）；为 spinner1 添加适配器。

⑥第 27~38 行，设置监听器。

spinner1. setOnItemSelectedListener( ) 方法主要是设置 spinner1 的下拉选中事件监听器 onItemSelected(AdapterView < ? > arg0，View arg1,**int** arg2，**long** arg3)。

其中 arg0 表示该适配器控件对象，arg1 表示选中的 View 对象，注意是指一个列表对象，arg2 表示选中的列表对象的下标，arg3 表示该 View 对象绑定的 id 值。

⑦第三步第 39~43 行，采用配置 array. xml 配置数据源，并且自定义弹出样式，修改 spinner 显示的样式。

6）图文 Spinner 应用

【例 3-2】 实现图 3.7 所示的自定义适配器界面。Spinner 显示国旗图文信息，选中某个条目，在文本框显示该国名称。

参照项目 1，在 Eclipse 中创建名为 Ex03_02 的工程。

（1）实现过程

①第一步。

复制 america. png、china. png、english. png、japan. png、korea. png 和 switzerland. png 到 res→drawable 目录。

在 res→values 目录下，新建 colors. xml。

< ? xml version = "1.0" encoding = "utf-8" ? >

< resources >

 < color name = "red" > #fd8d8d </color >

 < color name = "green" > #9cfda3 </color >

 < color name = "blue" > #8d9dfd </color >

 < color name = "white" > #FFFFFF </color >

 < color name = "black" > #000000 </color >

</resources >

图 3.7 自定义适配器界面

在 res→values 目录中，找到 strings. xml，新增几个字符串常量，文件内容如下：

< ? xml version = "1.0" encoding = "utf-8" ? >

< resources >

 < string name = "app_name" > Ex03_02 </string >

 < string name = "action_settings" > Settings </string >

 < string name = "hello_world" > Hello world! </string >

 < string name = "ys" > 选择的国家是 </string >

 < string name = "china" > 中国 </string >

 < string name = "english" > 英国 </string >

 < string name = "america" > 美国 </string >

 < string name = "japan" > 日本 </string >

 < string name = "korea" > 韩国 </string >

```
< string name = "switzerland" > 瑞士 </ string >
</ resources >
```

②第二步。

在 res→layout 下编辑 activity_main.xml，包含一个 TextView 和一个 Spinner。

```
< RelativeLayout xmlns:android = "http://schemas.android.com/apk/res/android"
    xmlns:tools = "http://schemas.android.com/tools"
    android:layout_width = "match_parent"
    android:layout_height = "match_parent"
    android:paddingBottom = "@dimen/activity_vertical_margin"
    android:paddingLeft = "@dimen/activity_horizontal_margin"
    android:paddingRight = "@dimen/activity_horizontal_margin"
    android:paddingTop = "@dimen/activity_vertical_margin"
    tools:context = ".MainActivity" >
    < TextView
        android:id = "@+id/textView1"
        android:layout_width = "wrap_content"
        android:layout_height = "wrap_content"
        android:text = "@string/ys"
        android:textSize = "24dip"
        />
    < Spinner
        android:id = "@+id/spinner1"
        android:layout_width = "wrap_content"
        android:layout_height = "wrap_content"
        android:layout_alignLeft = "@+id/textView1"
        android:layout_below = "@+id/textView1"
        android:layout_marginTop = "14dp" />
</ RelativeLayout >
```

③第三步。

编辑源文件 MainActivity.java。

1. **package** szpt.android.ex03_02;
2. **import** android.os.Bundle;
3. **import** android.app.Activity;
4. **import** android.view.Menu;
5. **import** android.view.View;
6. **import** android.view.ViewGroup;
7. **import** android.widget.AdapterView;
8. **import** android.widget.BaseAdapter;
9. **import** android.widget.Gallery;
10. **import** android.widget.ImageView;

11. **import** android. widget. LinearLayout;
12. **import** android. widget. Spinner;
13. **import** android. widget. TextView;
14. **import** android. widget. AdapterView. OnItemSelectedListener;
15. **public class** MainActivity **extends** Activity {
16.   //所有资源图片 id 的数组
17.   **int**[ ] images ={ R. drawable. *china*, R. drawable. *america*, R. drawable. *japan*, R. drawable. *korea*,
18.   R. drawable. *switzerland*};
19.   //所有资源字符串 id 的数组
20.   **int**[ ] *msgs* ={ R. string. *china*, R. string. *america*, R. string. *japan*, R. string. *korea*,
21.   R. string. *switzerland*};
22.   @Override
23.   **protected void** onCreate(Bundle savedInstanceState) {
24.   **super**. onCreate(savedInstanceState);
25.   setContentView(R. layout. *activity_main*);
26.   //初始化 Spinner
27.   Spinner sp =(Spinner) **this**. findViewById(R. id. *spinner1*);
28.   //为 Spinner 准备内容适配器
29.   BaseAdapter ba = **new** BaseAdapter( )
30.   {
31.   @Override
32.   **public int** getCount( ) {
33.   **return** *msgs*. length;//选项总个数
34.   }
35.   @Override
36.   **public** Object getItem(**int** arg0) { **return null**; }
37.   @Override
38.   **public long** getItemId(**int** arg0) { **return** 0; }
39.   @Override
40.   **public** View getView(**int** arg0, View arg1, ViewGroup arg2) {
41.   /*
42.   动态生成每个下拉项对应的 View,每个下拉项 View 由 LinearLayout
43.   中包含一个 ImageView 及一个 TextView 构成
44.   */
45.   //初始化 LinearLayout
46.   LinearLayout ll = **new** LinearLayout(MainActivity. **this**);
47.   ll. setOrientation(LinearLayout. *HORIZONTAL*);   //设置朝向
48.   //初始化 ImageView
49.   ImageView ii = **new** ImageView(MainActivity. **this**);
50.   ii. setImageDrawable(getResources( ). getDrawable(images[arg0]));//设置图片

```
51.            ii. setScaleType( ImageView. ScaleType. FIT_XY) ;//不按比例拉伸图片
52.            ii. setLayoutParams( new Gallery. LayoutParams( 60,60 ) ) ;
53.            ll. addView( ii ) ;                        //添加到 LinearLayout 中
54.            TextView tv = new TextView( MainActivity. this ) ;//初始化 TextView
55.            tv. setText( " " + getResources( ). getText( msgs[ arg0 ] ) ) ;//设置内容
56.            tv. setTextSize( 24 ) ;                    //设置字体大小
57.            tv. setTextColor( R. color. black ) ;      //设置字体颜色
58.            ll. addView( tv ) ;                        //添加到 LinearLayout 中
59.            return ll ;
60.        }
61.    } ;
62.    sp. setAdapter( ba ) ;//为 Spinner 设置内容适配器
63.    //设置选项选中的监听器
64.    sp. setOnItemSelectedListener(
65.    new OnItemSelectedListener( )
66.    {
67.        @Override
68.        public void onItemSelected( AdapterView <? > arg0, View arg1,
69.        int arg2, long arg3 ) {//重写选项被选中事件的处理方法
70.            TextView tv = ( TextView ) findViewById( R. id. textView1 ) ;//获取主界面 TextView
71.            LinearLayout ll = ( LinearLayout ) arg1 ;//获取当前选中选项对应的 LinearLayout
72.            TextView tvn = ( TextView ) ll. getChildAt( 1 ) ;//获取其中的 TextView
73.            StringBuilder sb = new StringBuilder( ) ;//用 StringBuilder 动态生成信息
74.            sb. append( getResources( ). getText( R. string. ys ) ) ;
75.            sb. append( " :" ) ;
76.            sb. append( tvn. getText( ) ) ;
77.            tv. setText( sb. toString( ) ) ;//信息设置进主界面 TextView
78.        }
79.        @Override
80.        public void onNothingSelected( AdapterView <? > arg0 ) { }
81.    }
82.    ) ;
83.    }
84.    @Override
85.    public boolean onCreateOptionsMenu( Menu menu ) {
86.        //Inflate the menu; this adds items to the action bar if it is present.
87.        getMenuInflater( ). inflate( R. menu. main, menu ) ;
88.        return true ;
89.    }
90. }
```

（2）代码分析

①第一步把程序所需资源图片文件资源复制到 res→drawable 目录中，并且新增了颜色常量和字符串常量，这些都会在 R 文件中产生索引号，在程序中可以通过 R 文件访问到需要的资源文件。

②第二步编辑 activity_main.xml，产生了一个 TextView 和 Spinner 控件。XML 文件访问资源文件是使用@String/方式，程序中访问资源文件是使用 R. 方式，注意两者的区别。

③第三步第 17~21 行，定义了两个整型数组，分别存放图片资源文件和字符串常量，作为 Spinner 的 Model 数据。

④第三步第 29~61 行，产生了一个 BaseAdapter 对象，可以看到产生一个 BaseAdapter 对象，必须覆盖以下方法：

public int getCount( )
public Object getItem(**int** arg0)
public long getItemId(**int** arg0)
public View getView(**int** arg0，View arg1，ViewGroup arg2)

这 4 个方法构成了一个适配器的必须的 4 个方法，其中 getCount( ) 方法提供 Model 的条数，getItem 和 getItemId 都是获取条目时候有用，而 getView( ) 方法主要是返回指定位置的 View 对象。这个是重点方法，因为这个方法会根据位置、Model 数据，产生一个 View 对象，填充到适配器控件中，其中 arg0 表示某个位置的 View 对象，因此根据此 arg0 获取 Model 中指定位置的图片和字符串，然后通过 Java 中创建对象的过程，创建了一个 LinearLayout 对象，包含一个 ImageView 对象和 TextView 对象，并调用相应的设置方法设置相关状态，最后返回的就是该 LinearLayout 对象。

⑤第三步第 62 行，将产生好的适配器对象设置到 Spinner 上，这样 Spinner 中的 View 对象的填充就由该适配器对象决定。

⑥第三步第 64~83 行，给 Spinner 控件添加了一个 ItemSelect 监听器对象，监听 Spinner 中选择条目的事件，可以看到 public void onItemSelected(AdapterView < ? > arg0，View arg1，int arg2，long arg3) 传入的参数 arg0 表示点击的 spinner1，arg1 表示选中的 spinner1 中的条目，注意这时候条目是一个 LinearLayout 对象，所以，如果想获取或者控制条目中的子控件，必须取出布局对象中的子控件，arg2 分别是选中条目的位置下标，arg3 则是该条目对应的 id。

3. ListView

①ListView 是列表视图，可以将一些零散的控件以列表的形式组织起来，并为其中的列表项添加事件监听。ListView 类层次如图 3.8 所示。

```
java.lang.Object
  ↳ android.view.View
      ↳ android.view.ViewGroup
          ↳ android.widget.AdapterView<T extends android.widget.Adapter>
              ↳ android.widget.AbsListView
                  ↳ android.widget.ListView
```

图 3.8　ListView 类层次

主要属性如下：
- android：divider 在列表中各个项之间的颜色或者颜色图片。
- android：dividerHeight 分隔条的高度。
- android：entries 填充 ListView 的数组资源。
- android：footerDividersEnabled 当被设置成 false 时，ListView 不将绘制底部 View 组件前的分隔条。
- android：headerDividersEnabled 当被设置成 false 时，ListView 不将绘制头部 View 组件后的分隔条。

②SimpleAdapter 主要用于在适配器控件中显示复杂的 View 对象，将集合对象中单个对象中的不同数据项填充到 View 中的不同组件中，并显示在适配器控件中的框架中。

- SimpleAdapter（Context context，List＜? extends Map＜String，?＞＞ data，int resource，String[ ] from，int[ ] to）

context：simpleAdapter 适配器相关联的上下文。
data：一个映射的列表．列表中的每一条对应 List 表中的一行。Maps 包含每行的数据。
resource：一行的样式文件。
from：Map 中的列名的列表。
to：对应的样式文件中的组件 ID 列表；这些应该都是 TextView 组件。

其中，from 和 to 之间是一一对应的关系，每个 Map 中的数据按照 from 和 to 的对应关系，将数据放入 resource 的对应位置，产生对应的 View 对象。

常用方法：
int　getCount（ ）：获得条目数量。
View　getDropDownView（int position，View convertView，ViewGroup parent）：返回数据集下拉列表中特定位置的 View 对象。
int　getPosition（T item）：返回 Model 数组中特定对象的位置。
Object　getItem（int position）：返回特定位置的 View 对象。
long　getItemId（int position）：返回特定位置 View 对象的绑定 ID 值。
View　getView（int position，View convertView，ViewGroup parent）：返回特定位置的 View 对象。
SimpleAdapter.ViewBinder　getViewBinder（ ）：返回 SimpleAdapter.ViewBinder 该类对象用于绑定数据到指定的 View 组件上。
void　setDropDownViewResource（int resource）：设置布局资源文件用于产生下拉 View。
void　setViewBinder（SimpleAdapter.ViewBinder viewBinder）：设置 binder 用于绑定数据到指定的 View 上。
void　setViewImage（ImageView v，int value）：用于设置 ImageView 上的图片，但是仅仅适用于不存在 ViewBinder 或者 ViewBinder 不起作用的情况。
void　setViewImage（ImageView v，String value）：用于设置 ImageView 上的图片，但是仅仅适用于不存在 ViewBinder 或者 ViewBinder 不起作用的情况。

void setViewText（TextView v，String text）：用于设置 TextView 上的文字，但是仅仅适用于不存在 ViewBinder 或者 ViewBinder 不起作用的情况。

③ListView 上的事件处理。

setOnClickListener(View. OnClickListener l)：当组件被点击时,注册一个监听器。

void setOnItemClickListener(AdapterView. OnItemClickListener listener)：当 ListView 中的某个 View 项被点击时,注册一个监听器。

void setOnItemLongClickListener（ AdapterView. OnItemLongClickListener listener）：当 ListView中的某个 View 项被长按时,注册一个监听器。

void setOnItemSelectedListener（ AdapterView. OnItemSelectedListener listener）：当 ListView中的某个 View 项被选中时,注册一个监听器。

【例3－3】 如图3.9 所示，以列表条目的方式显示多条 Model 数据。

参照项目1，在 Eclipse 中创建名为 Ex03_03 的工程。

（1） 实现过程

①第一步。

新建单行布局 personitem. xml，里面还有3 个 TextView，分别设置好相对位置、宽度和字体等相关信息。

<? xml version = "1. 0" encoding = "utf-8"? >
< RelativeLayout
　　xmlns：android = "http：//schemas. android. com/apk/res/android"
　　android：orientation = "vertical"
　　android：layout_width = "fill_parent"
　　android：layout_height = "fill_parent" >
　　　< TextView   android：id = "@ + id/personid" android：layout_width = "60px"   android：textSize = "25sp" android：he = "wrap_content"   >
　　　</TextView >
　　　< TextView android：layout_width = "160px" android：layout_height = "wrap_content" android：layout_toRightOf = "@id/personid" android：layout_alignTop = "@id/personid" android：gravity = "center_horizontal" android：id = "@ + id/name" >
　　　</TextView >
　　　< TextView android：layout_width = "wrap_content" android：layout_height = "wrap_content" android：layout_toRightOf = "@id/name" android：layout_alignTop = "@id/name" android：id = "@ + id/age" >
　　　</TextView >
</RelativeLayout >

②第二步。

修改 activity_main. xml 主界面布局文件，包含一个 ListView 适配器控件。

< RelativeLayout xmlns：android = "http：//schemas. android. com/apk/res/android"

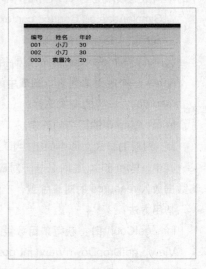

图3.9　ListView 多条目显示

```
        xmlns:tools = "http://schemas.android.com/tools"
        android:layout_width = "match_parent"
        android:layout_height = "match_parent"
        android:paddingBottom = "@dimen/activity_vertical_margin"
        android:paddingLeft = "@dimen/activity_horizontal_margin"
        android:paddingRight = "@dimen/activity_horizontal_margin"
        android:paddingTop = "@dimen/activity_vertical_margin"
        tools:context = ".MainActivity" >
        <ListView
            android:id = "@+id/listView1"
            android:layout_width = "match_parent"
            android:layout_height = "wrap_content"
            android:layout_alignParentLeft = "true" >
        </ListView>
</RelativeLayout>
```

③第三步。

修改 MainActivity.java 文件。

```
1.  package szpt.android.Ex03_03;
2.  import java.util.ArrayList;
3.  import java.util.HashMap;
4.  import java.util.List;
5.  import android.os.Bundle;
6.  import android.app.Activity;
7.  import android.util.Log;
8.  import android.view.Menu;
9.  import android.view.View;
10. import android.widget.AdapterView;
11. import android.widget.ListView;
12. import android.widget.SimpleAdapter;
13. import android.widget.AdapterView.OnItemSelectedListener;
14.
15. public class MainActivity extends Activity {
16.     ListView listView = null;
17.     List<HashMap<String, String>> data = null;
18.     SimpleAdapter adapter = null;
19.     @Override
20.     protected void onCreate(Bundle savedInstanceState) {
21.         super.onCreate(savedInstanceState);
22.         setContentView(R.layout.activity_main);
23.         listView = (ListView)findViewById(R.id.listView1);
```

```
24.    data = new ArrayList<HashMap<String,String>>();
25.    HashMap<String,String> title = new HashMap<String,String>();
26.    title.put("personid","编号");
27.    title.put("name","姓名");
28.    title.put("age","年龄");
29.    data.add(title);
30.    title = new HashMap<String,String>();
31.    title.put("personid","001");
32.    title.put("name","小刀");
33.    title.put("age","30");
34.    data.add(title);
35.    title = new HashMap<String,String>();
36.    title.put("personid","002");
37.    title.put("name","小刀");
38.    title.put("age","30");
39.    data.add(title);
40.    title = new HashMap<String,String>();
41.    title.put("personid","003");
42.    title.put("name","袁眉冷");
43.    title.put("age","20");
44.    data.add(title);
45.    adapter = new SimpleAdapter(MainActivity.this,
46.    data, R.layout.personitem, new String[]{"personid","name","age"},
47.    new int[]{R.id.personid, R.id.name, R.id.age});
48.    listView.setAdapter(adapter);
49.    listView.setOnItemSelectedListener(new OnItemSelectedListener(){
50.        public void onItemSelected(AdapterView<?> arg0, View arg1,
51.            int arg2, long arg3){
52.            // TODO Auto-generated method stub
53.            Log.v("personid", data.get(arg2).get("personid"));
54.            Log.v("name", data.get(arg2).get("name"));
55.            Log.v("age", data.get(arg2).get("age"));
56.        }
57.        public void onNothingSelected(AdapterView<?> arg0){
58.            // TODO Auto-generated method stub
59.        }
60.    });
61.    }
62.    @Override
63.    public boolean onCreateOptionsMenu(Menu menu){
```

```
64.        //Inflate the menu; this adds items to the action bar if it is present.
65.        getMenuInflater().inflate(R.menu.main, menu);
66.        return true;
67.    }
68. }
```

（2）代码分析

①第3步第16～18行，ListView 中 Model 数据的存储在 List < HashMap < String, String > > data 中，这个 data 是一个含有 HashMap 的集合对象，每个对象都是含有多个键值对，每个 HashMap 对应 Model 中的一条数据。

②第三步第24～44行，模拟产生了4条数据，每条数据用一个 HashMap 存储，多个 HashMap 放入一个 ArrayList 中，作为 ListView 的 Model 数据。

③第三步第45～47行，产生 ListView 的适配器对象

SimpleAdapter adapter = new SimpleAdapter(PersonActivity.this,
        data, R.layout.personitem, new String[]{"personid","name","age"},
        new int[]{R.id.personid, R.id.name, R.id.age});

其中 R.layout.personitem 表示 ListView 中某一行的布局文件。

new String[]{"personid","name","age"}参数表示取一个 HashMap 对应的键值。

new int[]{R.id.personid, R.id.name, R.id.age}表示 personitem 布局资源文件中，需要把 HashMap 中键对应的值填充到布局资源文件中的对应的组件 ID。

SimpleAdapter 中有这些参数以后，以 personitem 作为布局样式，依次取出 data 中的数据，按照 String 字符串数组和整型数组的对应关系，产生 View 对象，放入适配器控件中。

④第三步第48行，把适配器对象设置到控件上去。

⑤第三步第49～61行，为适配器控件中条目选中事件添加相应的处理器：

public void onItemSelected(AdapterView < ? > arg0, View arg1,**int** arg2, **long** arg3)

处理器方法中的 arg0 表示事件发生的适配器控件，arg1 表示选中的适配器控件中的那个条目 View 对象，arg2 表示所在位置的下标，arg3 表示条目对应的 id 值。

4. GridView

在 Android 开发中 GridView 是比较常用的组件，它以二维网格的形式展示具体内容，网格中的 View 对象来自于与 GridView 关联的 ListAdapter 对象。

常用方法：

- getAdapter()：返回 ListView 当前用的适配器。
- setAdapter（ListAdapter adapter）：设置这个 GridView 的适配器对象。
- setColumnWidth（int columnWidth）：设置 GridView 的列的宽度。
- setGravity（int gravity）：设置 GridView 中的 View 对象水平方向如何排列。
- setHorizontalSpacing（int horizontalSpacing）：设置 GridView 中每个 View 对象水平方向的间距。
- setNumColumns（int numColumns）：设置 GridView 的列的数量。

- setVerticalSpacing（int verticalSpacing）：设置 GridView 中各个 View 对象的垂直方向的间距。
- getCheckedItemPosition（）：返回当前被选中的项目，选择模式被设置为 CHOICE_MODE_SINGLE 时有效。复选框使用 getCheckedItemPositions（）方法。
- onKeyDown（int keyCode，KeyEvent event）：GridView 响应键盘按键事件，以及 onKeyUp（）、onTouchEvent（）等方法。

5. Gallery

Gallery 组件以居中、水平滚动列表方式显示数据项。Gallery 默认使用 Theme_gallery Item Background 作为 Gallery 中每个 View 组件的背景。如果不希望如此，需要调整 Gallery 属性，例如间隔距离等。Gallery 里面的组件应该使用 Gallery.LayoutParams 作为布局参数。在 Android 4.1（Android API level 16）之后，不再建议使用这个控件，但还对其有所支持。Android 4.1 中使用 HorizontableScroolView 和 ViewPager 可以达到同样效果。

```
java.lang.Object
    android.view.View
        android.view.ViewGroup
            android.widget.AdapterView<SpinnerAdapter>
                android.widget.AbsSpinner
                    android.widget.Gallery
```

6. TabHost

TabHost 是选项卡式窗口组件的容器。该组件拥有两个对象：tab 标签的集合和显示窗口的 FrameLayout 对象。容器对象控制显示在内部的窗口。TabHost 的使用有一些固定的格式。首先要求在布局文件的格式为 tabhost 标签里面添加 framelayout，在里面添加相应的控件，至少包括一个 framelayout 和 tabwidget，framelayout 必须命名为 @android：id/tabcontent，tabwidget 必须命名为 @android：id/tabs。这里，tabcontent 里面存放的是加载的多个 activity，tabs 里面存放的是与各个 activity 相对应的下面的按钮。需要注意的是，如果布局文件设定完毕，tabs 显示不出来，则需要在 tabcontent 里面设置 android：layout_weight="1" 即可。

```
java.lang.Object
    android.view.View
        android.view.ViewGroup
            android.widget.FrameLayout
                android.widget.TabHost
```

常用方法：
- void addTab(TabHost.TabSpec tabSpec)：添加 Tab 项。
- void clearAllTabs（）：移除这个 tabHost 关联的 TabWidget 中所有的 Tab 项。
- int getCurrentTab（）：获得当前 Tab 的下标值。
- String getCurrentTabTag（）：获得当前 Tab 的 Tag 字符串。
- View getCurrentTabView（）：获得当前 Tab 的 TabWidget 的 View 对象。

- View getCurrentView（ ）：获得当前 TabAtivity 中显示的 View 对象。
- FrameLayout getTabContentView（ ）：获得用来容纳 Tab 内容的 FrameLayout 对象。
- TabWidget getTabWidget（ ）：获得整个 TabWidget 标签栏。
- TabHost. TabSpec newTabSpec（String tag）：产生一个与 tabHost 关联的 TabHost. TabSpec 对象。
- void onTouchModeChanged（boolean isInTouchMode）：触摸模式改变的回调方法 。
- void setCurrentTab（int index）：用下标设置当前选中的 Tab。
- void setCurrentTabByTag（String tag）：用 tag 设置当前选中的 Tab。
- void setOnTabChangedListener（TabHost. OnTabChangeListener l）。

7. TabWidget

设置 Tab 状态发生变化的监听器对象

显示一系列 Tab 标签，用于代表 Tab 集合的每一页。这个部件的容器对象是 TabHost。当用户选择一个 Tab，这个对象发送一条消息给父容器 TabHost，让其转换需要显示的页面。一般不直接使用它的方法。容器 TabHost 被用来添加标签、添加和管理回调方法。这个对象主要用于遍历 Tab 对象，或者调整 Tab 列表的布局，但是大多数方法都是在 TabHost 对象中调用。

  java. lang. Object
    android. view. View
      android. view. ViewGroup
        android. widget. LinearLayout
          android. widget. TabWidget

常用方法：

- void addView(View child)：添加一个儿子 View 对象。
- void dispatchDraw(Canvas canvas)：被 draw（ ）方法调用绘制儿子 View 对象。
- void focusCurrentTab（int index）：设置当前 Tab 并且让当前 Tab 成为焦点。
- void onFocusChange（View v，boolean hasFocus）：当一个 view 对象的状态发生变化的时候被调用。
- void setCurrentTab（int index）：设置当前 Tab 。
- void setEnabled（boolean enabled）：设置这个 View 对象是否 Enabled。

8. TabActivity

TabActivity 是一种特殊的 Activity，用于包含多个内嵌的 Acitivities 或者 Views，所以它包含很多特定于多个 Activities 的方法。

  java. lang. Object
    android. content. Context
      android. content. ContextWrapper
        android. view. ContextThemeWrapper
          android. app. Activity
            android. app. ActivityGroup

android. app. TabActivity

常用方法：
- TabHost getTabHost( )：返回 Activity 加载的 TabHost 对象。
- TabWidget getTabWidget( )：返回 Activity 的 TabWidget，也就是标签栏。
- void onContentChanged( )：当 TabActivity 中包含的 Views 内容发生变化的时候，更新屏幕状态。
- void setDefaultTab(int index)：使用下标设置默认显示的 Tab 选项。
- void setDefaultTab(String tag)：使用 tag 设置默认的 Tab 选项。

任务1　显示商品列表

1. 任务说明

如图 3.10 所示，应用商店列表在 ListView 中显示多条 App 信息，每条 App 信息包含标题、图标、开发者、等级和状态。

图 3.10　商品列表显示

参照项目1，在 Eclipse 中创建名为 Ex03_04 的工程。

2. 实现过程
1）第一步
①编辑 strings.xml。

```
<?xml version="1.0" encoding="utf-8"?>
<resources>
    <string name="app_name">Ex03_04</string>
    <string name="action_settings">Settings</string>
    <string name="hello_world">Hello world!</string>
    <string name="download">下载</string>
</resources>
```

②复制 Ex03_04 所需要的图片文件和显示配置文件到 drawable-hdpi 下。

图片资源：baidusmall.jpg、breadsmall.jpg、cowsmall.jpg、market_star01.png、market_star02.png、market_star03.png、market_star04.png、market_star05.png、market_star06.png、market_star07.png、market_star08.png、market_star09.png、market_star10.png、progress.png、qqsmall.jpg、xiesmall.jpg、button_n.9.png、button_p.9.png 和 progress.png。

两个配置文件：button.xml 和 list_item_bg.xml。其中 button.xml 用来配置按钮背景不同状态下显示不同的图片；list_item_bg.xml 用来配置在什么状态下显示 progress.png。

button.xml

```xml
<?xml version="1.0" encoding="utf-8"?>
<selector xmlns:android="http://schemas.android.com/apk/res/android">
    <item android:state_focused="true"
        android:state_pressed="true"
        android:drawable="@drawable/button_p" />
    <item android:state_focused="false"
        android:state_pressed="true"
        android:drawable="@drawable/button_p" />
    <item android:state_focused="true"
        android:drawable="@drawable/button_p" />
    <item android:state_focused="false"
        android:drawable="@drawable/button_n" />
</selector>
```

list_item_bg.xml

```xml
<?xml version="1.0" encoding="utf-8"?>
<selector xmlns:android="http://schemas.android.com/apk/res/android">
    <item android:state_focused="true"
        android:state_pressed="true"
        android:drawable="@drawable/progress" />
    <item android:state_focused="false"
        android:state_pressed="true"
```

```
        android:drawable = "@drawable/progress" />
    <item android:state_focused = "true"
        android:drawable = "@drawable/progress" />
</selector>
```

③新建布局文件 featured.xml。

```
<RelativeLayout xmlns:android = "http://schemas.android.com/apk/res/android"
    android:layout_width = "match_parent"
    android:layout_height = "match_parent"
    android:background = "#ededed" >
        <ListView
            android:id = "@+id/featured_list"
            android:layout_width = "wrap_content"
            android:layout_height = "fill_parent"
            android:cacheColorHint = "#00000000"
            android:divider = "@null"
            android:focusable = "false"
            android:listSelector = "#00000000" />
</RelativeLayout>
```

④新建单行布局文件 app_list_item.xml。

```
<?xml version = "1.0" encoding = "utf-8"?>
<RelativeLayout xmlns:android = "http://schemas.android.com/apk/res/android"
    android:layout_width = "match_parent"
    android:layout_height = "match_parent"
    android:background = "@drawable/list_item_bg"
    >
        <LinearLayout
            android:layout_toLeftOf = "@+id/app_btn"
            android:layout_width = "fill_parent"
            android:layout_height = "wrap_content"
            android:orientation = "horizontal"
            android:layout_centerVertical = "true"
            >
            <ImageView
                android:layout_width = "50dp"
                android:layout_height = "50dp"
                android:layout_gravity = "center_vertical"
                android:id = "@+id/app_icon"
                android:src = "@drawable/ic_launcher"
                />
            <LinearLayout
```

```xml
        android:layout_marginLeft="10dp"
        android:layout_width="wrap_content"
        android:layout_height="wrap_content"
        android:orientation="vertical"
        >
            <TextView
                android:id="@+id/app_developer"
                android:layout_width="wrap_content"
                android:layout_height="wrap_content"
                android:text="developer"
                android:textColor="@android:color/darker_gray"
                android:textSize="10sp"
                android:lines="1"
                />
            <TextView
                android:id="@+id/app_name"
                android:layout_width="wrap_content"
                android:layout_height="wrap_content"
                android:text="name"
                android:textSize="12sp"
                />
        <ImageView
            android:id="@+id/app_star"
            android:layout_width="wrap_content"
            android:layout_height="wrap_content"
            android:src="@drawable/market_star01"
            />
    </LinearLayout>
</LinearLayout>
<Button
    android:id="@+id/app_btn"
    android:layout_width="wrap_content"
    android:layout_height="30dp"
    android:layout_alignParentRight="true"
    android:layout_centerVertical="true"
    android:text="download"
    android:textSize="10sp"
    android:paddingLeft="15dp"
    android:paddingRight="15dp"
    android:gravity="center_vertical"
```

```
            android:background = "@drawable/button"
            android:focusable = "false"
            / >
</RelativeLayout>
```

2) 第二步

显示下载的 App,针对 App 的信息,在包 szpt. android. ex03_04. beans 中新建一个 AppInfo. java,用于表示存储 App 的概况。

```
1.  package szpt.android.ex03_04.beans;
2.  public class AppInfo {
3.      private int id;
4.      private String name;
5.      private int downloadCount;
6.      private int star;
7.      private String developer;
8.      private int icon;
9.      private String apkFile;
10.     private String packge;
11.     private String categoryName;
12.     private String version;
13.     public AppInfo() {
14.     }
15.     public AppInfo(int id, String name, int star, String developer, int icon) {
16.         super();
17.         this.id = id;
18.         this.name = name;
19.         this.star = star;
20.         this.developer = developer;
21.         this.icon = icon;
22.     }
23.     public int getId() {
24.         return id;
25.     }
26.     public void setId(int id) {
27.         this.id = id;
28.     }
29.     public String getName() {
30.         return name;
31.     }
32.     public void setName(String name) {
33.         this.name = name;
```

```
34.     }
35.     public int getDownloadCount( ) {
36.         return downloadCount;
37.     }
38.     public void setDownloadCount( int downloadCount) {
39.         this. downloadCount = downloadCount;
40.     }
41.     public int getStar( ) {
42.         return star;
43.     }
44.     public void setStar( int star) {
45.         this. star = star;
46.     }
47.     public String getDeveloper( ) {
48.         return developer;
49.     }
50.     public void setDeveloper( String developer) {
51.         this. developer = developer;
52.     }
53.     public int getIcon( ) {
54.         return icon;
55.     }
56.     public void setIcon( int icon) {
57.         this. icon = icon;
58.     }
59.     public String getApkFile( ) {
60.         return apkFile;
61.     }
62.     public void setApkFile( String apkFile) {
63.         this. apkFile = apkFile;
64.     }
65.     public String getPackge( ) {
66.         return packge;
67.     }
68.     public void setPackge( String packge) {
69.         this. packge = packge;
70.     }
71.     public String getCategoryName( ) {
72.         return categoryName;
73.     }
```

```
74.     public void setCategoryName(String categoryName) {
75.         this.categoryName = categoryName;
76.     }
77.     public String getVersion() {
78.         return version;
79.     }
80.     public void setVersion(String version) {
81.         this.version = version;
82.     }
83. }
```

3) 第三步

在包 szpt.android.ex03_04.model 下,新建一个 ModelGenerator.java,用来模拟我们要产生的 Model 数据。

```
1. package szpt.android.ex03_04.model;
2. import java.util.ArrayList;
3. import szpt.android.ex03_04.R;
4. import szpt.android.ex03_04.beans.AppInfo;
5. public class ModelGenerator {
6.     //模拟 App 简要列表 Model 数据
7.     public static ArrayList<AppInfo> getFeatureApplyList() {
8.         ArrayList<AppInfo> applyList = new ArrayList<AppInfo>();
9.         AppInfo ai4 = new AppInfo(113, "百度旅游", 2, "Baidu",
10.            R.drawable.baidusmall);
11.        applyList.add(ai4);
12.        AppInfo ai3 = new AppInfo(108, "面包旅游", 3, "BreadTrip 公司",
13.            R.drawable.breadsmall);
14.        AppInfo ai5 = new AppInfo(115, "QQ", 2, "腾讯公司", R.drawable.qqsmall);
15.        applyList.add(ai5);
16.        AppInfo ai1 = new AppInfo(102, "途牛旅游", 2, "途牛公司", R.drawable.cowsmall);
17.        applyList.add(ai1);
18.        AppInfo ai2 = new AppInfo(105, "携程旅游", 2, "携程公司", R.drawable.xiesmall);
19.        applyList.add(ai2);
20.        applyList.add(ai3);
21.        return applyList;
22.    }
23. }
```

4) 第四步

在包 szpt.android.ex03_04.adapter 下新建一个自定义适配器 ListItemAdapter.java。

```
1. package szpt.android.ex03_04.adapter;
2. import java.util.List;
```

```
3.  import szpt.android.ex03_04.R;
4.  import szpt.android.ex03_04.beans.AppInfo;
5.  import android.content.Context;
6.  import android.view.LayoutInflater;
7.  import android.view.View;
8.  import android.view.ViewGroup;
9.  import android.widget.BaseAdapter;
10. import android.widget.Button;
11. import android.widget.ImageView;
12. import android.widget.TextView;
13. public class ListItemAdapter extends BaseAdapter {
14.     private List < AppInfo > mAppList;
15.     private LayoutInflater mInflater;
16.     private Context mContext;
17.     private class RecentViewHolder {
18.         private TextView list_title;
19.         private TextView list_Author;
20.         private ImageView list_start;
21.         private ImageView list_icon;
22.         private Button list_download;
23.     }
24.     public ListItemAdapter( Context c, List < AppInfo > mAppList) {
25.         this.mContext = c;
26.         this.mAppList = mAppList;
27.         mInflater = ( LayoutInflater ) c
28.                 .getSystemService( Context.LAYOUT_INFLATER_SERVICE );
29.     }
30.     public List < AppInfo > getmAppList( ) {
31.         return mAppList;
32.     }
33.     public void setmAppList( List < AppInfo > mAppList) {
34.         this.mAppList = mAppList;
35.     }
36.     public void clear( ) {
37.         if ( mAppList ! = null ) {
38.             mAppList.clear( );
39.         }
40.     }
41.     @Override
42.     public int getCount( ) {
```

```
43.        return mAppList.size();
44.    }
45.    public Object getItem(int position) {
46.        return mAppList.get(position);
47.    }
48.    public long getItemId(int position) {
49.        // TODO Auto-generated method stub
50.        return position;
51.    }
52.    public View getView(final int position, View convertView, ViewGroup parent) {
53.
54.        final RecentViewHolder holder;
55.        if (convertView == null) {
56.            convertView = mInflater.inflate(R.layout.app_list_item, null);
57.            holder = new RecentViewHolder();
58.            holder.list_title = (TextView) convertView
59.                    .findViewById(R.id.app_name);
60.            holder.list_Author = (TextView) convertView
61.                    .findViewById(R.id.app_developer);
62.            holder.list_start = (ImageView) convertView
63.                    .findViewById(R.id.app_star);
64.            holder.list_icon = (ImageView) convertView
65.                    .findViewById(R.id.app_icon);
66.            holder.list_download = (Button) convertView
67.                    .findViewById(R.id.app_btn);
68.            convertView.setTag(holder);
69.        } else {
70.            holder = (RecentViewHolder) convertView.getTag();
71.        }
72.        final String title = mAppList.get(position).getName();
73.        holder.list_title.setText(title);
74.        String author = mAppList.get(position).getDeveloper();
75.        holder.list_Author.setText(author);
76.        int appState = mAppList.get(position).getStar();
77.        //设置 list_download 设置成"下载"
78.        holder.list_download.setText(mContext.getString(R.string.download));
79.        holder.list_download.setEnabled(true);
80.        holder.list_start.setBackgroundResource(R.drawable.market_star03);
81.        holder.list_icon.setImageResource(mAppList.get(position).getIcon());
82.        return convertView;
```

83.   }
84. }

5） 第五步

在包 szpt.android.ex03_04 包下新建 Feature.java，代码如下：

1. **package** szpt.android.ex03_04;
2. **import** szpt.android.ex03_04.adapter.ListItemAdapter;
3. **import** szpt.android.ex03_04.model.ModelGenerator;
4. **import** android.app.Activity;
5. **import** android.os.Bundle;
6. **import** android.view.View;
7. **import** android.widget.AdapterView;
8. **import** android.widget.AdapterView.OnItemClickListener;
9. **import** android.widget.ListView;
10. **import** android.widget.RelativeLayout;
11. **import** android.widget.TextView;
12. **import** android.widget.Toast;
13. **public class** Feature **extends** Activity {
14.   ListView listView;
15.   @Override
16.   **protected void** onCreate(Bundle savedInstanceState) {
17.   **super**.onCreate(savedInstanceState);
18.   setContentView(R.layout.*featured*);
19.   listView = (ListView)findViewById(R.id.*featured_list*);
20.   ListItemAdapter lia = **new** ListItemAdapter(Feature.**this**, ModelGenerator.getFeature*ApplyList*());
21.   listView.setAdapter(lia);
22.   listView.setOnItemClickListener(**new** OnItemClickListener() {
23.   @Override
24.   **public void** onItemClick(AdapterView<?> arg0, View arg1, **int** arg2,
25.   **long** arg3) {
26.   //**TODO** Auto-generated method stub
27.   RelativeLayout ll = (RelativeLayout)arg1;//获取当前选中选项对应的 RelativeLayout
28.   TextView tv = (TextView)arg1.findViewById(R.id.*app_name*);
29.   StringBuilder sb = **new** StringBuilder();//用 StringBuilder 动态生成信息
30.   sb.append(tv.getText());
31.   sb.append("被点击了!");
32.   String stemp = sb.toString();
33.   Toast.makeText(Feature.**this**, stemp, Toast.LENGTH_LONG).show();
34.   }

35.   });
36.   }
37. }

6）第六步

在配置文件 AndroidManifest.xml 中，把启动 <activity> 标签中的 android：name 替换成 Feature。

<activity
        android：name = "szpt.android.ex03_04.Feature"
        android：label = "@string/app_name" >
    <intent-filter >
        <action android：name = "android.intent.action.MAIN" />
        <category android：name = "android.intent.category.LAUNCHER" / >
    </intent-filter >
</activity >

3. 代码分析

①第二步新建的 AppInfo.java 中，包含一个应用程序所有相关的信息，使用这个 Bean 对象存储一个 App 信息。

②第三步 ModelGenerator.java，主要是根据资源文件数据，模拟产生一些 App Bean 对象，存储到一个 ArrayList 中，作为 ListView 的 Model 数据。

③第四步。

- 第 53～83 行，是自定义 BaseAapter 的关键方法，和 Spinner 中的自定义适配器原理相同，自定义适配器 BaseAdapter，主要是根据 Model 数据和单行布局文件，返回多个 View 对象填充到 ListView 控件中。
- 第 54～71 行，使用单行布局文件 app_list_item.xml 产生指定的 View 对象，并利用已有 convertView 对象，其中的 RecentViewHolder 类主要是使用原有的 View 对象，避免重现产生浪费资源。
- 第 71～83 行，从 Model mAppList 中取出一条数据，填充到 holder 中，注意 holder 和 convertView 是绑定的，最后返回产生好的 convertView 对象。

④第五步。

- 第 18～21 行，加载布局文件，获取 ListView 控件，然后生成适配器，加载到 ListView 控件上。
- 第 22～34 行，产生一个匿名监听器对象，监听 ListView 上的条目点击事件：
  public void onItemClick（AdapterView <？> arg0，View arg1，**int** arg2，**long** arg3）
  响应方法中的 arg0 表示点击的适配器控件，arg1 表示点击条目的 View 对象，arg2 表示该 View 对象在适配器控件中的位置，arg3 对应点击条目的 View 对象的 id 值。
  在这个方法实现过程中，将 arg1 下溯造型回原来单行布局的容器类型，然后取出该容器中的组件，获取其文本内容，用 Toast 显示在界面上。

任务 2 显示分类商品

1. 任务说明

该项目使用 ListView 和 GridView 混合使用的方法,界面效果如图 3.11 所示,可以看到 ListView 里面每个条目的内容由 ImageView + GridView 构成,ImageView 用来显示分类名称,GridView 用来显示该分类下的 App 的简要内容。

参照项目 1,在 Eclipse 中创建名为 Ex03_06 的工程。

如图 3.12 所示,任务需要 3 个布局文件,分别为 classify_page.xml、classify_list_item.xml 和 app_grid_item.xml,其中 classify_page.xml 为页面布局文件,还有一个 ListView,而 classify_list_item.xml 含有一个 ImageView 和 GridView,布局文件定义了 classify_page.xml 中 ListView 中的一个条目的布局,也就是 ListView 中的每个条目都是按照 classify_list_item.xml 定义的布局进行显示,classify_list_item.xml 中 GridView 的每个项的显示方式定义在 app_grid_item.xml 中。

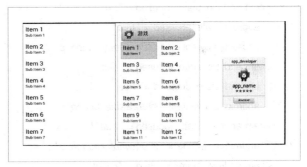

图 3.11 分类商品显示　　　　　　图 3.12 ListView 布局文件

自定义适配器 BaseAdapter 要注意,ListView 定义的自定义适配器返回的 View 对象,是 classify_list_item.xml 定义的布局对象,在生成 View 对象的时候,要同时给 GridView 定义一个自定义适配器对象。

2. 实现过程

1)第一步

①编辑 strings.xml。

<? xml version = "1.0" encoding = "utf-8"? >

< resources >

```xml
<string name = "app_name" > Ex03_06 </string >
<string name = "action_settings" > Settings </string >
<string name = "hello_world" > Hello world! </string >
<string name = "download" > 下载 </string >
<string name = "ok" > 确定 </string >
<string name = "cancel" > 取消 </string >
</resources >
```

②复制 Ex03_06 所需要的图片文件和显示配置文件到 drawable – hdpi 下。

图片资源：market_star01.png、market_star02.png、market_star03.png、market_star04.png、market_star05.png、market_star06.png、market_star07.png、market_star08.png、market_star09.png、input_n.9.png、footlikeright_n.png、footlikeright_p.png、market_star10.png、banana.png、bridgeme.png、facebook.png、flypigg.png、qq.png、traffic.png、twitter.png、webchat.png、category_game1、category_social chat.png、facebook.png、button_n.9.png、button_p.9.png、gridapp_bg_n.png 和 gridapp_bg_p.png。

3 个配置文件：button.xml、gridapp_bg.xml 和 footlikeright.xml，其中 button.xml 用来配置按钮背景不同状态下显示不同的图片；gridapp_bg.xml 用来配置 GridView 中的 View 对象不同状态下的背景，footlikeright.xml 主要用来定义分类显示部分在不同状态下的背景图片。

button.xml

```xml
<? xml version = "1.0" encoding = "utf-8"? >
< selector xmlns:android = "http://schemas.android.com/apk/res/android" >
    < item android:state_focused = "true"
        android:state_pressed = "true"
        android:drawable = "@drawable/button_p" / >
    < item android:state_focused = "false"
        android:state_pressed = "true"
        android:drawable = "@drawable/button_p" / >
    < item android:state_focused = "true"
        android:drawable = "@drawable/button_p" / >
    < item android:state_focused = "false"
        android:drawable = "@drawable/button_n" / >
</ selector >
```

gridapp_bg.xml

```xml
<? xml version = "1.0" encoding = "utf-8"? >
< selector xmlns:android = "http://schemas.android.com/apk/res/android" >
    < item android:state_focused = "true"
        android:state_pressed = "true"
        android:drawable = "@drawable/gridapp_bg_p" / >
    < item android:state_focused = "false"
        android:state_pressed = "true"
```

android:drawable = "@drawable/gridapp_bg_p"/>
    <item android:state_focused = "true"
        android:drawable = "@drawable/gridapp_bg_p"/>
    <item android:state_focused = "false"
        android:drawable = "@drawable/gridapp_bg_n"/>
</selector>

footlikeright.xml

<?xml version = "1.0" encoding = "utf-8"?>
<selector xmlns:android = "http://schemas.android.com/apk/res/android">
<item android:state_focused = "true"
    android:state_pressed = "true"
    android:drawable = "@drawable/footlikeright_p"/>
<item android:state_focused = "false"
    android:state_pressed = "true"
    android:drawable = "@drawable/footlikeright_p"/>
<item android:state_focused = "true"
    android:drawable = "@drawable/footlikeright_p"/>
<item android:state_focused = "false"
    android:drawable = "@drawable/footlikeright_n"/>
</selector>

③新建布局文件 classify_page.xml。

```
<?xml version = "1.0" encoding = "utf-8"?>
<LinearLayout xmlns:android = "http://schemas.android.com/apk/res/android"
    android:layout_width = "match_parent"
    android:layout_height = "match_parent"
    android:orientation = "vertical">
    <ListView
        android:id = "@+id/classify_list"
        android:layout_width = "fill_parent"
        android:layout_height = "fill_parent"
        >
    </ListView>
</LinearLayout>
```

④新建 ListView 中单行布局 classify_list_item.xml。

```
<?xml version = "1.0" encoding = "utf-8"?>
<LinearLayout xmlns:android = "http://schemas.android.com/apk/res/android"
    android:layout_width = "match_parent"
    android:layout_height = "match_parent"
    android:orientation = "vertical"
    >
```

```xml
<LinearLayout
    android:layout_width = "fill_parent"
    android:layout_height = "fill_parent"
    android:orientation = "vertical"
    android:layout_margin = "5dp"
    android:background = "@drawable/input_n"
    >
    <LinearLayout
        android:id = "@+id/classify_title"
        android:layout_width = "fill_parent"
        android:layout_height = "60dp"
        android:orientation = "horizontal"
        android:background = "@drawable/footlikeright"
        android:paddingBottom = "5dp"
        android:paddingRight = "40dp"
        android:layout_gravity = "center_vertical"
        android:clickable = "true"
        >
        <ImageView
            android:id = "@+id/classify_icon"
            android:layout_width = "45dp"
            android:layout_height = "45dp"
            android:layout_marginLeft = "15dp"
            android:src = "@drawable/ic_launcher"
            android:layout_gravity = "center_vertical"
            />
        <TextView
            android:id = "@+id/classify_name"
            android:layout_width = "wrap_content"
            android:layout_height = "wrap_content"
            android:layout_gravity = "center_vertical"
            android:layout_marginLeft = "10dp"
            android:text = "游戏"
            android:textSize = "20sp" />
    </LinearLayout>
    <GridView
        android:id = "@+id/classift_list_item_grid"
        android:layout_width = "fill_parent"
        android:layout_height = "400dp"
        android:numColumns = "2"
```

```
            android:paddingLeft = "7dp"
            >
        </GridView>
    </LinearLayout>
</LinearLayout>
```

⑤新建单行布局文件 app_grid_item.xml。

```
<?xml version = "1.0" encoding = "utf-8"?>
<LinearLayout xmlns:android = "http://schemas.android.com/apk/res/android"
    android:layout_width = "match_parent"
    android:layout_height = "match_parent"
    android:orientation = "vertical"
    android:gravity = "center"
    >
    <LinearLayout
        android:layout_width = "wrap_content"
        android:layout_height = "wrap_content"
        android:background = "@drawable/gridapp_bg"
        android:layout_margin = "5dp"
        android:orientation = "vertical"
        >
        <TextView
            android:id = "@+id/app_developer"
            android:layout_width = "fill_parent"
            android:layout_height = "wrap_content"
            android:gravity = "center"
            android:layout_marginTop = "5dp"
            android:text = "app_developer"
            android:textSize = "15sp"
            android:lines = "1"
            android:paddingLeft = "1dp"
            android:paddingRight = "1dp"
            />
        <ImageView
            android:id = "@+id/app_icon"
            android:layout_width = "fill_parent"
            android:layout_height = "60dp"
            android:gravity = "center"
            android:layout_marginTop = "10dp"
            android:src = "@drawable/ic_launcher"
            />
```

```xml
<TextView
    android:id = "@+id/app_name"
    android:layout_width = "fill_parent"
    android:layout_height = "wrap_content"
    android:gravity = "center"
    android:text = "app_name"
    android:textSize = "20sp"
    android:lines = "1"
    android:paddingLeft = "1dp"
    android:paddingRight = "1dp"
    />
<ImageView
    android:id = "@+id/app_star"
    android:layout_width = "fill_parent"
    android:layout_height = "wrap_content"
    android:gravity = "center"
    android:src = "@drawable/market_star07"
    />
<Button
    android:id = "@+id/app_btn"
    android:layout_width = "90dp"
    android:layout_height = "30dp"
    android:layout_marginTop = "12dp"
    android:layout_gravity = "center_horizontal"
    android:background = "@drawable/button"
    android:text = "@string/download"
    android:textSize = "10sp"
    android:focusable = "false"
    />
        </LinearLayout>
    </LinearLayout>
```

2）第二步

显示下载的 App，针对 App 的信息，在包 szpt.android.ex03_06.beans 新建两个 Beans：AppInfo.java，用于表示 App 的概要信息，Category.java 用于表示 App 分类信息。

AppInfo.java

1. **package** szpt.android.ex03_06.beans;
2. **public class** AppInfo {
3.   **private int** id;
4.   **private** String name;
5.   **private int** downloadCount;

```
6.      private int star;
7.      private String developer;
8.      private int icon;
9.      private String apkFile;
10.     private String packge;
11.     private String categoryName;
12.     private String version;
13.     public AppInfo( ) {
14.     }
15.     public AppInfo(int id, String name, int star, String developer, int icon) {
16.         super( );
17.         this. id = id;
18.         this. name = name;
19.         this. star = star;
20.         this. developer = developer;
21.         this. icon = icon;
22.     }
23.     public int getId( ) {
24.         return id;
25.     }
26.     public void setId(int id) {
27.         this. id = id;
28.     }
29.     public String getName( ) {
30.         return name;
31.     }
32.     public void setName(String name) {
33.         this. name = name;
34.     }
35.     public int getDownloadCount( ) {
36.         return downloadCount;
37.     }
38.     public void setDownloadCount(int downloadCount) {
39.         this. downloadCount = downloadCount;
40.     }
41.     public int getStar( ) {
42.         return star;
43.     }
44.     public void setStar(int star) {
45.         this. star = star;
```

```
46.    }
47.    public String getDeveloper( ) {
48.        return developer;
49.    }
50.    public void setDeveloper(String developer) {
51.        this.developer = developer;
52.    }
53.    public int getIcon( ) {
54.        return icon;
55.    }
56.    public void setIcon(int icon) {
57.        this.icon = icon;
58.    }
59.    public String getApkFile( ) {
60.        return apkFile;
61.    }
62.    public void setApkFile(String apkFile) {
63.        this.apkFile = apkFile;
64.    }
65.    public String getPackge( ) {
66.        return packge;
67.    }
68.    public void setPackge(String packge) {
69.        this.packge = packge;
70.    }
71.    public String getCategoryName( ) {
72.        return categoryName;
73.    }
74.    public void setCategoryName(String categoryName) {
75.        this.categoryName = categoryName;
76.    }
77.    public String getVersion( ) {
78.        return version;
79.    }
80.    public void setVersion(String version) {
81.        this.version = version;
82.    }
83. }
```

Category.java

```
1. package szpt.android.ex03_06.beans;
```

```
2.  public class Category {
3.      private int id;
4.      private String name;
5.      private int icon;
6.      public Category(int id, String name, int icon) {
7.          super();
8.          this.id = id;
9.          this.name = name;
10.         this.icon = icon;
11.     }
12.     public int getId() {
13.         return id;
14.     }
15.     public void setId(int id) {
16.         this.id = id;
17.     }
18.     public String getName() {
19.         return name;
20.     }
21.     public void setName(String name) {
22.         this.name = name;
23.     }
24.     public int getIcon() {
25.         return icon;
26.     }
27.     public void setIcon(int icon) {
28.         this.icon = icon;
29.     }
30. }
```

3）第三步

在包 szpt.android.ex03_06.model 下，新建一个 ModelGenerator.java，用来模拟要产生的 Model 数据，因为是显示分类下的信息，有两个方法：一个用于产生分类信息，一个用于产生分类下的 App 列表信息。

```
1.  package szpt.android.ex03_06.model;
2.  import java.util.ArrayList;
3.  import szpt.android.ex03_06.R;
4.  import szpt.android.ex03_06.beans.AppInfo;
5.  import szpt.android.ex03_06.beans.Category;
6.  import android.content.Context;
7.  import android.graphics.drawable.Drawable;
```

```java
8.  public class ModelGenerator {
9.  //获得分类列表
10.     public static ArrayList<Category> getCategoryList() {
11.         ArrayList<Category> categoryList = new ArrayList<Category>();
12.         Category temp1 = new Category(101, "游戏", R.drawable.category_game1);
13.         categoryList.add(temp1);
14.         Category temp2 = new Category(105, "社交与聊天",
15.             R.drawable.category_socialchat);
16.         categoryList.add(temp2);
17.         return categoryList;
18.     }
19.     //获得 App 简要列表
20.     public static ArrayList<AppInfo> getCategoryAppList(String categoryName) {
21.         ArrayList<AppInfo> applyList = new ArrayList<AppInfo>();
22.         if(categoryName.equals("游戏")) {
23.             AppInfo ai1 = new AppInfo(113, "Fly Pig", 2, "ZhiYuan Group",
24.                 R.drawable.flypigg);
25.             AppInfo ai2 = new AppInfo(108, "Banana", 3, "FDG 娱乐",
26.                 R.drawable.banana);
27.             AppInfo ai3 = new AppInfo(115, "赛车", 2, "Blitzblaster 软件公司",
28.                 R.drawable.traffic);
29.             AppInfo ai4 = new AppInfo(102, "BridgeMe", 2, "Snagon 工作室",
30.                 R.drawable.bridgeme);
31.             ai1.setCategoryName("游戏");
32.             ai2.setCategoryName("游戏");
33.             ai3.setCategoryName("游戏");
34.             ai4.setCategoryName("游戏");
35.             applyList.add(ai1);
36.             applyList.add(ai2);
37.             applyList.add(ai3);
38.             applyList.add(ai4);
39.         } else if (categoryName.equals("社交与聊天")) {
40.             AppInfo ai1 = new AppInfo(113, "推特", 2, "Twitter 公司",
41.                 R.drawable.twitter);
42.             AppInfo ai2 = new AppInfo(108, "微信", 3, "腾讯公司",
43.                 R.drawable.webchat);
44.             AppInfo ai3 = new AppInfo(115, "QQ", 2, "腾讯公司", R.drawable.qq);
45.             AppInfo ai4 = new AppInfo(102, "脸谱", 2, "FaceBook 公司",
46.                 R.drawable.facebook);
47.             ai1.setCategoryName("社交与聊天");
```

```
48.        ai2.setCategoryName("社交与聊天");
49.        ai3.setCategoryName("社交与聊天");
50.        ai4.setCategoryName("社交与聊天");
51.        applyList.add(ai1);
52.        applyList.add(ai2);
53.        applyList.add(ai3);
54.        applyList.add(ai4);
55.     }
56.     return applyList;
57. }
```

4）第四步

在包 szpt.android.ex03_06.tools 下，新建一个 ApplyListConvertListItem.java 提供一个静态方法 getListeItem，用来把 AppInfo 的 ArrayList 转换成适合适配器使用的 HashMap 的 ArrayList。

```
1.  package szpt.android.ex03_06.tools;
2.  import java.util.ArrayList;
3.  import java.util.HashMap;
4.  import szpt.android.ex03_06.R;
5.  import szpt.android.ex03_06.beans.AppInfo;
6.  import android.content.Context;
7.  public class ApplyListConvertListItem {
8.      private static int[] stars = {
9.          R.drawable.market_star01,
10.         R.drawable.market_star02, R.drawable.market_star03,
11.         R.drawable.market_star04, R.drawable.market_star05,
12.         R.drawable.market_star06, R.drawable.market_star07,
13.         R.drawable.market_star08, R.drawable.market_star09,
14.         R.drawable.market_star10 };
15.     public static ArrayList<HashMap<String,Object>>
16.         getListItem(Context ctx,ArrayList<AppInfo> applyList,ArrayList<HashMap<String,Object>> listItem){
17.         if(listItem == null){
18.             listItem = new ArrayList<HashMap<String,Object>>();
19.         }
20.         for(AppInfo temp:applyList){
21.             HashMap<String,Object> hm = new HashMap<String,Object>();
22.             hm.put("Icon", temp.getIcon());
23.             hm.put("Developer", temp.getDeveloper());
24.             hm.put("Name", temp.getName());
25.             hm.put("Star", stars[temp.getStar()-1]);
```

```
26.        hm.put("id", temp.getId());
27.        hm.put("APKFile", temp.getApkFile());
28.        hm.put("Packge", temp.getPackge());
29.        hm.put("Version", temp.getVersion());
30.        listItem.add(hm);
31.    }
32.    return listItem;
33. }
34. }
```

5）第五步

在包 szpt.android.ex03_06.adapter 下，新建一个 GridItemSimpleAdapter.java，用来给 GridView 填充 View 对象。

```
1.  package szpt.android.ex03_06.adapter;
2.  import java.util.List;
3.  import java.util.Map;
4.  import szpt.android.ex03_06.R;
5.  import android.content.Context;
6.  import android.graphics.Bitmap;
7.  import android.graphics.drawable.Drawable;
8.  import android.view.View;
9.  import android.view.ViewGroup;
10. import android.widget.Button;
11. import android.widget.ImageView;
12. import android.widget.SimpleAdapter;
13. public class GridItemSimpleAdapter extends SimpleAdapter {
14.     private List<? extends Map<String, ?>> data;
15.     private Context context;
16.     public GridItemSimpleAdapter(Context context,
17.         List<? extends Map<String, ?>> data, int resource, String[] from,
18.         int[] to) {
19.         super(context, data, resource, from, to);
20.         this.data = data;
21.         this.setViewBinder(new ViewBinder() {
22.             @Override
23.             public boolean setViewValue(View view, Object data,
24.                 String textRepresentation) {
25.                 if ((view instanceof ImageView) & ((data instanceof Bitmap)
26.                     || data instanceof Drawable) {
27.                     ImageView iv = (ImageView) view;
28.                     if (data instanceof Bitmap) {
```

```
29.            Bitmap bm = (Bitmap) data;
30.            setImageBitmap(bm);
31.         } else {
32.            Drawable drawable = (Drawable) data;
33.            iv.setImageDrawable(drawable);
34.         }
35.         return true;
36.      }
37.      return false;
38.    }
39.   });
40.   this.context = context;
41. }
42. @Override
43. public View getView(int position, View convertView, ViewGroup parent) {
44.    //TODO Auto-generated method stub
45.    convertView = super.getView(position, convertView, parent);
46.    try {
47.       Button appOperate = (Button) convertView.findViewById(R.id.app_btn);
48.       appOperate.setText(context.getResources().getString(R.string.download));
49.    } catch (Exception e) {
50.    }
51.    return convertView;
52.  }
53. }
```

6) 第六步

在包szpt.android.ex03_06.listener包下,新建一个AppListItemClickListener类,用于监听GridView上的ItemClick事件。

```
1. package szpt.android.ex03_06.listeners;
2. import java.util.ArrayList;
3. import java.util.HashMap;
4. import szpt.android.ex03_06.R;
5. import android.content.Context;
6. import android.view.View;
7. import android.widget.AdapterView;
8. import android.widget.AdapterView.OnItemClickListener;
9. import android.widget.LinearLayout;
10. import android.widget.TextView;
11. import android.widget.Toast;
12. public class AppListItemClickListener implements OnItemClickListener {
```

13. private Context ctx;
14. ArrayList<HashMap<String, Object>> listItem;
15. public AppListItemClickListener(Context ctx, ArrayList<HashMap<String, Object>> listItem) {
16.   super();
17.   this.ctx = ctx;
18.   this.listItem = listItem;
19. }
20. @Override
21. public void onItemClick(AdapterView<?> arg0, View arg1, int arg2, long arg3) {
22.   LinearLayout ag = (LinearLayout)arg1;
23.   TextView tv = (TextView)ag.findViewById(R.id.*app_name*);
24.   Toast.makeText(ctx,"应用程序:"+tv.getText().toString()+"被选中!",Toast.LENGTH_LONG).show();
25. }
26. }

7）第七步

在包 szpt.android.ex03_06 包下新建 ClassifyListAdapter.java，用来给 ListView 进行 View 对象的填充。

1. **package** szpt.android.ex03_06.adapter;
2. **import** java.util.ArrayList;
3. **import** java.util.HashMap;
4. **import** szpt.android.ex03_06.ClassifyActivity;
5. **import** szpt.android.ex03_06.R;
6. **import** szpt.android.ex03_06.beans.AppInfo;
7. **import** szpt.android.ex03_06.beans.Category;
8. **import** szpt.android.ex03_06.listeners.AppListItemClickListener;
9. **import** szpt.android.ex03_06.model.ModelGenerator;
10. **import** szpt.android.ex03_06.tools.ApplyListConvertListItem;
11. **import** android.content.Context;
12. **import** android.graphics.Bitmap;
13. **import** android.graphics.drawable.BitmapDrawable;
14. **import** android.graphics.drawable.Drawable;
15. **import** android.os.Handler;
16. **import** android.view.LayoutInflater;
17. **import** android.view.View;
18. **import** android.view.View.OnClickListener;
19. **import** android.view.ViewGroup;
20. **import** android.widget.BaseAdapter;
21. **import** android.widget.GridView;
22. **import** android.widget.ImageView;

```
23.    import android.widget.LinearLayout;
24.    import android.widget.TextView;
25.    import android.widget.Toast;
26.    public class ClassifyListAdapter extends BaseAdapter{
27.        private Context ctx;
28.        private ArrayList<Category> list;
29.        private Handler h;
30.        public ClassifyListAdapter(Context ctx,ArrayList<Category> list,Handler h){
31.            this.ctx = ctx;
32.            this.list = list;
33.            this.h = h;
34.        }
35.        @Override
36.        public int getCount(){
37.            //TODO Auto-generated method stub
38.            return this.list.size();
39.        }
40.        @Override
41.        public Object getItem(int arg0){
42.            //TODO Auto-generated method stub
43.            return this.list.get(arg0);
44.        }
45.        @Override
46.        public long getItemId(int arg0){
47.            //TODO Auto-generated method stub
48.            return arg0;
49.        }
50.        @Override
51.        public View getView(int arg0,View arg1,ViewGroup arg2){
52.            //TODO Auto-generated method stub
53.            Category category = list.get(arg0);
54.            LinearLayout ll = (LinearLayout) LayoutInflater.from(ctx).
55.                inflate(R.layout.classify_list_item,null);
56.            LinearLayout classify_title = (LinearLayout) ll.findViewById(R.id.classify_title);
57.            ImageView classify_icon = (ImageView) ll.findViewById(R.id.classify_icon);
58.            TextView classify_name = (TextView) ll.findViewById(R.id.classify_name);
59.            GridView classift_list_item_grid =
60.                (GridView) ll.findViewById(R.id.classift_list_item_grid);
61.            Drawable icon = ctx.getResources().getDrawable(category.getIcon());
62.            classify_icon.setImageDrawable(icon);
```

```
63.     classify_name.setText(category.getName());
64.     //获得分类下详细App信息
65.     ArrayList<AppInfo>
        applyList = ModelGenerator.getCategoryAppList(category.getName());
66.     ArrayList<HashMap<String,Object>> listItem =
67.     ApplyListConvertListItem.getListItem(ctx, applyList, null);
68.     GridItemSimpleAdapter sa = new GridItemSimpleAdapter(ctx, listItem, R.layout.app_grid_item,
69.     new String[]{"Icon","Developer","Name","Star"},
70.     new int[]{R.id.app_icon,R.id.app_developer,R.id.app_name,R.id.app_star});
71.     classift_list_item_grid.setAdapter(sa);
72.     classift_list_item_grid.setOnItemClickListener(new AppListItemClickListener(ctx,listItem));
73.     classify_title.setOnClickListener(new MyClassifyTitleClickListener(ctx,category,icon));
74.     return ll;
75.     }
76.     private class MyClassifyTitleClickListener implements OnClickListener{
77.         private Context ctx;
78.         private Category category;
79.         private Bitmap icon;
80.         public MyClassifyTitleClickListener(Context ctx,Category category,Drawable icon2){
81.             this.ctx = ctx;
82.             this.category = category;
83.             this.icon = ((BitmapDrawable)icon2).getBitmap();
84.         }
85.         @Override
86.         public void onClick(View arg0){
87.             Toast.makeText(ctx,"分类:"+category.getName()+"被选中了",Toast.LENGTH_LONG).show();
88.         }
89.     }
90. }
```

8）第八步

编写主Activity文件ClassifyActivity.java。这是程序的主界面，用来显示分类App信息。

```
1. package szpt.android.ex03_06;
2. import java.util.ArrayList;
3. import szpt.android.ex03_06.adapter.ClassifyListAdapter;
4. import szpt.android.ex03_06.beans.Category;
5. import szpt.android.ex03_06.model.ModelGenerator;
6. import android.app.Activity;
```

```java
7.  import android.app.AlertDialog;
8.  import android.app.Dialog;
9.  import android.app.DialogFragment;
10. import android.app.AlertDialog.Builder;
11. import android.content.DialogInterface;
12. import android.os.Bundle;
13. import android.os.Handler;
14. import android.os.Message;
15. import android.view.KeyEvent;
16. import android.widget.ListView;
17. public class ClassifyActivity extends Activity{
18.     private ListView classify_list;
19.     private ArrayList<Category> categoryList;
20.     ClassifyListAdapter ba;
21.     @Override
22.     protected void onCreate(Bundle savedInstanceState){
23.         // TODO Auto-generated method stub
24.         super.onCreate(savedInstanceState);
25.         setContentView(R.layout.classify_page);
26.         classify_list = (ListView)findViewById(R.id.classify_list);
27.         initCategoryList();
28.         ba = new ClassifyListAdapter(this,categoryList,h);
29.         classify_list.setAdapter(ba);
30.     }
31.     private void initCategoryList(){
32.         categoryList = ModelGenerator.getCategoryList();
33.     }
34.     @Override
35.     public boolean onKeyDown(int keyCode, KeyEvent event){
36.         // TODO Auto-generated method stub
37.         if(keyCode == KeyEvent.KEYCODE_BACK && event.getRepeatCount() == 0){
38.             showEveryDialog(0);
39.             return true;
40.         }else
41.             return super.onKeyDown(keyCode, event);
42.     }
43.     /*
44.     该对话框的实现已经被deprecated,需要使用DialogFragment替换
45.     */
```

```
46.    void showEveryDialog(int id){
47.        DialogFragment df = MyAlertDialogFragment.newInstance(id);
48.        df.show(getFragmentManager(),"dialog");
49.    }
50.    public static class MyAlertDialogFragment extends DialogFragment{
51.        public static MyAlertDialogFragment newInstance(int id){
52.            MyAlertDialogFragment frag = new MyAlertDialogFragment();
53.            Bundle args = new Bundle();
54.            args.putInt("id", id);
55.            frag.setArguments(args);
56.            return frag;
57.        }
58.        /**
59.         覆写Fragment类的onCreateDialog()方法,在FragmentDialog的show(;)方法执行之后,系统
              会调用这个回调方法。
60.         */
61.        @Override
62.        public Dialog onCreateDialog(Bundle saveInstanceState){
63.            // 获取对象实例化时传入的窗口标题
64.            int id = getArguments().getInt("id");
65.            Dialog dialog = null;
66.            Builder b = new AlertDialog.Builder(getActivity());
67.            switch(id){
68.                case 0:
69.                    b.setMessage("您确定要退出?");
70.                    b.setNegativeButton(R.string.ok,
71.                        new DialogInterface.OnClickListener(){
72.                            @Override
73.                            public void onClick(DialogInterface arg0, int arg1){
74.                                // TODO Auto-generated method stub
75.                                // FileDownloadList.stopAllFileDownload();
76.                                System.exit(0);
77.                            }
78.                    });
79.                    b.setPositiveButton(R.string.cancel, null);
80.                    dialog = b.create();
81.                    break;
82.            }
```

```
83.            return dialog;
84.        }
85.    }
86. }
```

9）第九步

在配置文件 AndroidManifest.xml 中，把启动 < activity > 标签中的 android：name 替换成 ClassifyActivity。

```
< activity
        android：name = " szpt. android. ex03_06. ClassifyActivity"
        android：label = " @ string/app_name" >
    < intent – filter >
            < action android：name = " android. intent. action. MAIN" / >
            < category android：name = " android. intent. category. LAUNCHER" / >
    </ intent – filter >
</ activity >
```

3. 代码分析

①第一步主要是准备好项目所需要的图片文件、字符串常量的配置和 View 所需要的组件图片的配置文件，以及项目所需要的布局文件。因为这个项目是 ListView 和 GridView 混合使用，所以包含三个布局文件，其中 classify_page. xml 是主界面的布局文件，包含一个 ListView，classify_list_item. xml 是 ListView 中单个条目的布局文件，其中包含一个 GridView 控件，app_grid_item. xml 则是 GridView 中单个项目的布局文件。

②第二步准备好两个 beans，分别表示分类的信息和应用程序的简要信息，AppInfo 类中的 categoryName 表示应用程序属于具体分类。

③第三步新建的 ModelGenerator 类提供了两个方法，分别模拟产生分类数据和分类下的应用程序信息，其中 getCategoryList( ) 方法产生"游戏"和"社交与聊天"两个分类，并设置相应的信息，getCategoryAppList(String categoryName) 分别根据参数 categoryName 的值，产生相应分类下的应用程序列表。

④第四步新建的 ApplyListConvertListItem 作用是把 AppInfo 的 ArrayList 转换成适合适配器使用的 HashMap 的 ArrayList，因为在表示应用程序信息时，使用 AppInfo 的 Bean 对象存储一个应用程序的信息，而适配器对象的作用是将 Map 的数据和行布局组件进行匹配，所以需要把 AppInfo 的 Bean 对象转换成 HashMap 对象。

⑤第五步新建的 GridItemSimpleAdapter 作用是给混合布局中的 GridView 提供适配器。前面给 GridView 填充 View 对象的时候，使用的是自定义的 BaseAdapter，重写了 BaseAdapter 的几个关键方法，在这里并没有使用自定义的 BaseAdapter，而是采用继承 SimpleAdapter 的方式。SimpleAdapter 是 Android 专门用来提供给单行多条数据进行 View 填充的适配器对象，只要给定行布局，Model 数据和相关的 Model 数据到行布局的映射关系就能够自动进行 View 对象的填充。SimpleAdapter 也是一种自定义的 BaseAdapter，大家可以查看其

源码，对于 SimpleAdapter 的行为会有更为深入的了解，同时可以学习自定义适配器的编写。

- 第 16～41 行，在 GridItemSimpleAdapter 构造器方法中，进行了参数的传递，同时指定了一个 ViewBinder 对象，该对象的作用是指定 Map 数据和行布局中的具体组件如何进行匹配。SimpleAdapter 的实现过程中，已经能够匹配大多数数据，但有些数据还是不能进行匹配，如 Bitmap 或 Drawable 对象匹配到 ImageView 上没有实现，构造器中指定的 ViewBinder 对象扩展了 SimpleAdapter 类的功能，扩展了 ImageView 能够匹配的数据。ViewBinder 会在 getView( ) 方法之前起作用，如果不能匹配，则使用 getView( ) 方法进行匹配。
- 第 43～52 行重写了 SimpleAdapter 的 getView( ) 方法，先调用 SimpleAdapter 的默认实现进行匹配，然后把按钮的文本数据设置成"下载"。

⑥第六步新建的 AppListItemClickListener 监听器类主要用于监听 GridView 上的单个应用程序的点击事件，以 Toast 方式显示具体某个应用程序被点击了。

⑦第七步新建的 ClassifyListAdapter 采用自定义适配器 BaseAdapter 的方式。要注意的是，getView( ) 方法返回的 View 对象本身也是一个带适配器控件的 View。

- 第 51～75 行重写了 BaseAdapter 的 getView( ) 方法，获取 Category 对象，然后产生 ListView 下的单行布局文件对象，取 Model 数据，设置相应数据，然后根据分类名称获取分类下的 App 的 Model 数据，设置当行布局中 GridView 的适配器对象。最后分别对分类和单个 App 监听。
- 第 76～89 行，"分类"点击的事件监听内部类，直接用 Toast 显示分类信息。

⑧第八步主 Activity – ClassifyActivity，程序启动 Activity。

- 第 24～33 行，加载布局文件，设置 ListView 的适配器对象。
- 第 46～86 行，定义一个回退提示对话框，使用前面学的 DialogFragment。
- 第 35～42 行，监听按键事件，如果按键是回退按钮，则显示提示退化框，提示是否退出程序。

任务 3　显示商品详情

1. 任务说明

本项目实现如图 3.13 所示。运行程序后，第一个界面是 Ex03_06 工程中已经编写好的，使用 ListView + GridView 混编而成。上面的工程完成之后，当选中某个应用时，只是以 Toast 在界面上显示某某应用选中了，本节修改上面的工程，当选中某个 App 时，能够进行 Activity 的跳转，在第二个界面显示 App 的详细信息，包括名称、图片、大小、开发商、产品描述和 App 的预览界面。

在第二个界面中，App 的预览界面使用 Gallery 进行展示。

参照项目 1，在 Eclipse 中打开 Ex03_07 的起始工程，该工程和上节中的 Ex03_06 工程一样,下面在该工程基础上完成应用程序详细信息的展示。

图 3.13 商品详情显示

2. 实现过程

1) 第一步

①编辑 strings.xml,新增 Description 常量,存储"描述"字符串常量。

```
<?xml version="1.0" encoding="utf-8"?>
<resources>
    <string name="app_name">Ex03_07</string>
    <string name="action_settings">Settings</string>
    <string name="hello_world">Hello world!</string>
    <string name="download">下载</string>
    <string name="ok">确定</string>
    <string name="cancel">取消</string>
    <string name="Description">描述</string>
</resources>
```

②在 res→drawable-hdpi 下新增 Ex03_07 所需要的图片文件和显示配置文件。

图片资源:category_fashionshopping.png、ebay.png、ebaypreview01.jpg、ebaypreview02.jpg、ebaypreview03.jpg、ebaypreview04.jpg、facebook.png、screen_view_seekpoint_highlight.png、screen_view_seekpoint_normal.png、tip_icon.png。

③参照图 3.14 所示设计界面和控件布局方式,新建布局文件 classify_page.xml。

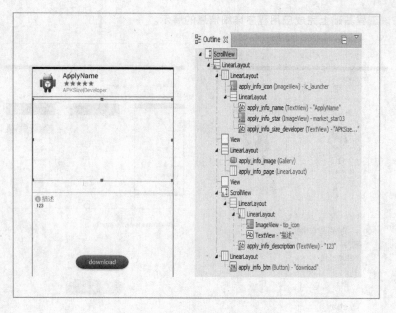

图3.14　商品详情布局

```xml
<?xml version = "1.0" encoding = "utf-8"?>
<ScrollView    xmlns:android = "http://schemas.android.com/apk/res/android"
    android:layout_width = "match_parent"
    android:layout_height = "match_parent"
    >
<LinearLayout
    android:layout_width = "match_parent"
    android:layout_height = "match_parent"
    android:orientation = "vertical"   >
    <LinearLayout
        android:layout_width = "fill_parent"
        android:layout_height = "wrap_content"
        android:orientation = "horizontal"
        android:layout_marginTop = "5dp"
        >
        <ImageView
            android:id = "@+id/apply_info_icon"
            android:layout_width = "50dp"
            android:layout_height = "50dp"
            android:layout_marginLeft = "10dp"
            android:layout_marginRight = "10dp"
            android:src = "@drawable/ic_launcher"
            />
```

```xml
<LinearLayout
    android:layout_width="fill_parent"
    android:layout_height="50dp"
    android:orientation="vertical"
    >
    <TextView
        android:id="@+id/apply_info_name"
        android:layout_width="wrap_content"
        android:layout_height="wrap_content"
        android:text="ApplyName"
        android:textSize="15sp"
        />
    <ImageView
        android:id="@+id/apply_info_star"
        android:layout_width="wrap_content"
        android:layout_height="wrap_content"
        android:src="@drawable/market_star03"
        />
    <TextView
        android:id="@+id/apply_info_size_developer"
        android:layout_width="wrap_content"
        android:layout_height="wrap_content"
        android:text="APKSize|Developer"
        android:textColor="@android:color/darker_gray"
        android:textSize="12sp"
        android:lines="1"
        />
</LinearLayout>
</LinearLayout>
<View
    android:layout_marginTop="5dp"
    android:layout_marginBottom="5dp"
    android:layout_width="fill_parent"
    android:layout_height="1.5dp"
    android:background="@android:color/darker_gray"
    />
<LinearLayout
    android:layout_height="200dp"
    android:layout_width="fill_parent"
    android:orientation="vertical"
```

```xml
        >
        <Gallery
            android:id = "@+id/apply_info_image"
            android:layout_height = "180dp"
            android:layout_width = "fill_parent"
            />
        <LinearLayout
            android:id = "@+id/apply_info_page"
            android:layout_width = "wrap_content"
            android:layout_height = "wrap_content"
            android:orientation = "horizontal"
            android:layout_gravity = "center_horizontal"
            >
        </LinearLayout>
    </LinearLayout>
    <View
        android:layout_marginTop = "5dp"
        android:layout_marginBottom = "5dp"
        android:layout_width = "fill_parent"
        android:layout_height = "1.5dp"
        android:background = "@android:color/darker_gray"
        />
    <ScrollView
        android:layout_marginLeft = "3dp"
        android:layout_marginRight = "3dp"
        android:layout_width = "fill_parent"
        android:layout_height = "120dp"
        >
        <LinearLayout
        android:layout_width = "fill_parent"
        android:layout_height = "wrap_content"
        android:orientation = "vertical"
        >
            <LinearLayout
                android:layout_width = "wrap_content"
                android:layout_height = "wrap_content"
                android:orientation = "horizontal"
                >
                <ImageView
                    android:layout_width = "wrap_content"
```

```xml
                    android:layout_height = "13dp"
                    android:src = "@drawable/tip_icon"
                    android:layout_gravity = "center_vertical"
                    />
                <TextView
                    android:layout_width = "wrap_content"
                    android:layout_height = "wrap_content"
                    android:text = "@string/Description"
                    android:textColor = "@android:color/darker_gray"
                    android:layout_gravity = "center_vertical"
                    />
            </LinearLayout>
            <TextView
                android:id = "@+id/apply_info_description"
                android:layout_width = "fill_parent"
                android:layout_height = "wrap_content"
                android:layout_marginLeft = "5dp"
                android:layout_marginRight = "5dp"
                android:text = "123"
                android:textSize = "10sp"
                />
        </LinearLayout>
    </ScrollView>
    <LinearLayout
        android:layout_width = "wrap_content"
        android:layout_height = "wrap_content"
        android:orientation = "horizontal"
        android:layout_gravity = "center_horizontal|bottom"
        android:layout_marginBottom = "10dp"
        android:layout_marginTop = "10dp"
        >
        <Button
            android:id = "@+id/apply_info_btn"
            android:layout_width = "120dp"
            android:layout_height = "40dp"
            android:text = "download"
            android:textColor = "@android:color/white"
            android:textSize = "14sp"
            android:background = "@drawable/install"
            android:layout_marginLeft = "20dp"
```

```
                    android:layout_marginRight = "20dp"
                    />
            </LinearLayout>
        </LinearLayout>
    </ScrollView>
```

2）第二步

第二个界面显示 App 的详细信息，在包 szpt.android.ex03_07.beans 中新增一个 Bean。AppInfoDetails.java，用于表示 App 的详细信息。

AppInfoDetails.java

1. **package** szpt.android.ex03_07.beans;
2. **import** java.io.Serializable;
3. **import** java.util.ArrayList;
4. **import** android.graphics.drawable.Drawable;
5. //App 的详细信息
6. public class AppInfoDetails **implements** Serializable{
7.   **private int** id;
8.   **private** String name;
9.   **private** String categoryName;
10.   **private int** star;
11.   **private** String introduce;
12.   **private** String version;
13.   **private** String packageName;
14.   **private** String updateTime;
15.   **private** String developer;
16.   **private** String apkFile;
17.   **private int** downloadCount;
18.   **private int** apkSize;
19.   ArrayList < Drawable > permissionList;
20.   **public** AppInfoDetails(**int** id, String name, String categoryName, **int** star,
21.   String introduce, String version, String packageName,
22.   String updateTime, String developer, String apkFile,
23.   **int** downloadCount, **int** apkSize){
24.   super();
25.   **this**.id = id;
26.   **this**.name = name;
27.   **this**.categoryName = categoryName;
28.   **this**.star = star;
29.   **this**.introduce = introduce;
30.   **this**.version = version;
31.   **this**.packageName = packageName;

```
32.         this.updateTime = updateTime;
33.         this.developer = developer;
34.         this.apkFile = apkFile;
35.         this.downloadCount = downloadCount;
36.         this.apkSize = apkSize;
37.     }
38.     public int getId() {
39.         return id;
40.     }
41.     public void setId(int id) {
42.         this.id = id;
43.     }
44.     public String getName() {
45.         return name;
46.     }
47.     public void setName(String name) {
48.         this.name = name;
49.     }
50.     public String getCategoryName() {
51.         return categoryName;
52.     }
53.     public void setCategoryName(String categoryName) {
54.         this.categoryName = categoryName;
55.     }
56.     public int getStar() {
57.         return star;
58.     }
59.     public void setStar(int star) {
60.         this.star = star;
61.     }
62.     public String getIntroduce() {
63.         return introduce;
64.     }
65.     public void setIntroduce(String introduce) {
66.         this.introduce = introduce;
67.     }
68.     public String getVersion() {
69.         return version;
70.     }
71.     public void setVersion(String version) {
```

```
72.         this.version = version;
73.     }
74.     public String getPackageName() {
75.         return packageName;
76.     }
77.     public void setPackageName(String packageName) {
78.         this.packageName = packageName;
79.     }
80.     public String getUpdateTime() {
81.         return updateTime;
82.     }
83.     public void setUpdateTime(String updateTime) {
84.         this.updateTime = updateTime;
85.     }
86.     public String getDeveloper() {
87.         return developer;
88.     }
89.     public void setDeveloper(String developer) {
90.         this.developer = developer;
91.     }
92.     public String getApkFile() {
93.         return apkFile;
94.     }
95.     public void setApkFile(String apkFile) {
96.         this.apkFile = apkFile;
97.     }
98.     public int getDownloadCount() {
99.         return downloadCount;
100.    }
101.    public void setDownloadCount(int downloadCount) {
102.        this.downloadCount = downloadCount;
103.    }
104.    public int getApkSize() {
105.        return apkSize;
106.    }
107.    public void setApkSize(int apkSize) {
108.        this.apkSize = apkSize;
109.    }
110.    public ArrayList<Drawable> getPermissionList() {
111.        return permissionList;
```

```
112.        }
113.        public void setPermissionList(ArrayList<Drawable> permissionList){
114.            this.permissionList = permissionList;
115.        }
116.    }
```

3) 第三步

在包 szpt.android.ex03_07.model 下修改 ModelGenerator.java，其中 getCategoryList() 方法中新增加一个分类"时尚与购物"，getCategoryAppList(String categoryName) 方法中新增加"时尚与购物"分类下的 App 信息，该分类下只添加了一个 ebay 的 App 简要信息，新增 AppInfoDetails getApplyInfo(Context ctx) 方法用于模拟一个 App 的详细信息。

```
1.  package szpt.android.ex03_07.model;
2.  import java.util.ArrayList;
3.  import szpt.android.ex03_07.R;
4.  import szpt.android.ex03_07.beans.AppInfo;
5.  import szpt.android.ex03_07.beans.AppInfoDetails;
6.  import szpt.android.ex03_07.beans.Category;
7.  import android.content.Context;
8.  import android.graphics.drawable.Drawable;
9.  public class ModelGenerator {
10.     //获得分类列表
11.     public static ArrayList<Category> getCategoryList(){
12.         ArrayList<Category> categoryList = new ArrayList<Category>();
13.         Category temp1 = new Category(101,"游戏",R.drawable.category_game1);
14.         categoryList.add(temp1);
15.         Category temp2 = new Category(105,"社交与聊天",
16.             R.drawable.category_socialchat);
17.         categoryList.add(temp2);
18.         Category temp3 = new Category(106,"时尚与购物",
19.             R.drawable.category_fashionshopping);
20.         categoryList.add(temp3);
21.         return categoryList;
22.     }
23.     //获得App简要列表
24.     public static ArrayList<AppInfo> getCategoryAppList(String categoryName){
25.         ArrayList<AppInfo> applyList = new ArrayList<AppInfo>();
26.         if(categoryName.equals("游戏")){
27.             AppInfo ai1 = new AppInfo(113,"Fly Pig",2,"ZhiYuan Group",
28.                 R.drawable.flypigg);
29.             AppInfo ai2 = new AppInfo(108,"Banana",3,"FDG娱乐",
30.                 R.drawable.banana);
```

```
31.        AppInfo ai3 = new AppInfo(115, "赛车", 2, "Blitzblaster 软件公司",
32.        R. drawable. traffic);
33.        AppInfo ai4 = new AppInfo(102, "BridgeMe", 2, "Snagon 工作室",
34.        R. drawable. bridgeme);
35.        ai1. setCategoryName("游戏");
36.        ai2. setCategoryName("游戏");
37.        ai3. setCategoryName("游戏");
38.        ai4. setCategoryName("游戏");
39.        applyList. add(ai1);
40.        applyList. add(ai2);
41.        applyList. add(ai3);
42.        applyList. add(ai4);
43.    } else if (categoryName. equals("社交与聊天")) {
44.        AppInfo ai1 = new AppInfo(113, "推特", 2, "Twitter 公司",
45.        R. drawable. twitter);
46.        AppInfo ai2 = new AppInfo(108, "微信", 3, "腾讯公司",
47.        R. drawable. webchat);
48.        AppInfo ai3 = new AppInfo(115, "QQ", 2, "腾讯公司", R. drawable. qq);
49.        AppInfo ai4 = new AppInfo(102, "脸谱", 2, "FaceBook 公司",
50.        R. drawable. facebook);
51.        ai1. setCategoryName("社交与聊天");
52.        ai2. setCategoryName("社交与聊天");
53.        ai3. setCategoryName("社交与聊天");
54.        ai4. setCategoryName("社交与聊天");
55.        applyList. add(ai1);
56.        applyList. add(ai2);
57.        applyList. add(ai3);
58.        applyList. add(ai4);
59.    } else if (categoryName. equals("时尚与购物")) {
60.        AppInfo ai1 = new AppInfo(120, "ebay", 2, "EBAY 公司",
61.        R. drawable. ebay);
62.        ai1. setCategoryName("时尚与购物");
63.        applyList. add(ai1);
64.    }
65.    return applyList;
66. }
67. //获得应用详细信息
68. public static AppInfoDetails getApplyInfo(Context ctx) {
69.    AppInfoDetails applyInfo = new AppInfoDetails(102,"eBay","时尚和购物",
70.    3,"著名的在线购物网站 eBay 的官方客户端!"
```

71. ,"2.5"," com. ebay. mobile"," 2014 – 02 – 26 17:09:30"," eBay Mobile"," ebay. apk",111,
785);
72. //初始化预览图片
73. ArrayList < Drawable > imageList = **new** ArrayList < Drawable > ( );
74. Drawable imageBm1 = ctx. getResources( )
75. . getDrawable(R. drawable. *ebaypreview01*);
76. Drawable imageBm2 = ctx. getResources( )
77. . getDrawable(R. drawable. *ebaypreview02*);
78. Drawable imageBm3 = ctx. getResources( )
79. . getDrawable(R. drawable. *ebaypreview03*);
80. Drawable imageBm4 = ctx. getResources( )
81. . getDrawable(R. drawable. *ebaypreview04*);
82. Drawable imageBm5 = ctx. getResources( )
83. . getDrawable(R. drawable. *ebaypreview05*);
84. imageList. add(imageBm1);
85. imageList. add(imageBm2);
86. imageList. add(imageBm3);
87. imageList. add(imageBm4);
88. imageList. add(imageBm5);
89. applyInfo. setPermissionList(imageList);
90. **return** applyInfo;
91. }
92. }

4) 第四步

在包 szpt. android. ex03_07 下新建第二个界面对应的 Activity，ApplyInfoActivity. java，用于显示 App 的详细信息。

1. **package** szpt. android. ex03_07;
2. **import** java. util. ArrayList;
3. **import** szpt. android. ex03_07. beans. AppInfoDetails;
4. **import** szpt. android. ex03_07. model. ModelGenerator;
5. **import** android. app. Activity;
6. **import** android. content. Context;
7. **import** android. graphics. drawable. Drawable;
8. **import** android. os. Bundle;
9. **import** android. view. View;
10. **import** android. view. ViewGroup;
11. **import** android. widget. AdapterView;
12. **import** android. widget. AdapterView. OnItemSelectedListener;
13. **import** android. widget. BaseAdapter;
14. **import** android. widget. Button;

```
15. import android.widget.Gallery
16. import android.widget.ImageView;
17. import android.widget.LinearLayout;
18. import android.widget.TextView;
19. public class ApplyInfoActivity extends Activity {
20.     private int id;
21.     private int icon;
22.     private AppInfoDetails applyInfo;
23.     private ArrayList<Drawable> imageList;
24.     private ImageView apply_info_icon, apply_info_star;
25.     private TextView apply_info_name, apply_info_size_developer,
26.         apply_info_description;
27.     private Button apply_info_btn;
28.     private Gallery apply_info_image;
29.     private LinearLayout apply_info_page;
30.     @Override
31.     protected void onCreate(Bundle savedInstanceState) {
32.         //TODO Auto-generated method stub
33.         super.onCreate(savedInstanceState);
34.         setContentView(R.layout.apply_info_page);
35.         Bundle extras = getIntent().getExtras();
36.         id = extras.getInt("id");
37.         icon = extras.getInt("icon");
38.         initViews();
39.         initAppInfoDetails();
40.         initInterface();
41.         apply_info_icon.setImageResource(icon);
42.     }
43.     private void initViews() {
44.         apply_info_icon = (ImageView)findViewById(R.id.apply_info_icon);
45.         apply_info_star = (ImageView)findViewById(R.id.apply_info_star);
46.         apply_info_name = (TextView)findViewById(R.id.apply_info_name);
47.         apply_info_size_developer = (TextView)findViewById(R.id.apply_info_size_developer);
48.         apply_info_description = (TextView)findViewById(R.id.apply_info_description);
49.         // apply_info_jurisdiction = (Button)findViewById(R.id.apply_info_jurisdiction);
50.         apply_info_btn = (Button)findViewById(R.id.apply_info_btn);
51.         // apply_info_share = (Button)findViewById(R.id.apply_info_share);
52.         apply_info_image = (Gallery)findViewById(R.id.apply_info_image);
53.         apply_info_page = (LinearLayout)findViewById(R.id.apply_info_page);
54.     }
```

```
55.  private void initAppInfoDetails( ) {
56.      //初始化一个 App 详细信息
57.      this. applyInfo = ModelGenerator. getApplyInfo( this ) ;
58.      this. imageList = this. applyInfo. getPermissionList( ) ;
59.  }
60.  private void initInterface( ) {
61.      apply_info_icon. setImageResource( icon ) ;
62.      apply_info_name. setText( applyInfo. getName( ) ) ;
63.      int[ ] stars = {  R. drawable. market_star01,
         R. drawable. market_star02, R. drawable. market_star03, R. drawable. market_star04,
64.      R. drawable. market _ star05,  R. drawable. market _ star06,  R. drawable. market _ star07,
         R. drawable. market_star08, R. drawable. market_star09, R. drawable. market_star10 } ;
65.      apply_info_star. setImageResource( stars[ applyInfo. getStar( ) ] ) ;
66.      float size = applyInfo. getApkSize( ) / ( float ) 1024f / ( float ) 1024f;
67.      size = ( float )  ( Math. round( size * 100 ) ) / 100;
68.      apply_info_size_developer. setText( size + " MB | " + applyInfo. getDeveloper( ) ) ;
69.      apply_info_description. setText( applyInfo. getIntroduce( ) ) ;
70.      apply_info_image. setAdapter( new ImageAdapater( this, this. imageList ) ) ;
71.      initImagePage( ) ;
72.      apply_info_image. setOnItemSelectedListener( new OnItemSelectedListener( ) {
73.          @Override
74.          public void onItemSelected( AdapterView < ? > arg0, View arg1,
75.          int arg2, long arg3 ) {
76.              //TODO Auto - generated method stub
77.              for ( int i = 0; i < apply_info_page. getChildCount( ) ; i + + ) {
78.              ImageView iv = ( ImageView ) apply_info_page. getChildAt( i ) ;
79.              if ( i = = arg2 ) {
80.                  iv. setImageResource( R. drawable. screen_view_seekpoint_highlight ) ;
81.              } else {
82.                  iv. setImageResource( R. drawable. screen_view_seekpoint_normal ) ;
83.              }
84.              }
85.          }
86.          @Override
87.          public void onNothingSelected( AdapterView < ? > arg0 ) {
88.              //TODO Auto - generated method stub
89.          }
90.      } ) ;
91.  }
```

```
 92.    private void initImagePage( ) {
 93.       for ( int i = 0; i < imageList. size( ); i + + ) {
 94.          ImageView iv = new ImageView( this) ;
 95.          if ( i = = 0) {
 96.             iv. setImageResource( R. drawable. screen_view_seekpoint_highlight) ;
 97.          } else {
 98.             iv. setImageResource( R. drawable. screen_view_seekpoint_normal) ;
 99.          }
100.          iv. setLayoutParams( new Gallery. LayoutParams(
101.          LinearLayout. LayoutParams. WRAP_CONTENT ,LinearLayout. LayoutParams. WRAP_CONTENT ) ) ;
102.          apply_info_page. addView( iv) ;
103.       }
104.    }
105.    private class ImageAdapater extends BaseAdapter {
106.       private ArrayList < Drawable > imageList;
107.       private Context ctx;
108.       public ImageAdapater( Context ctx, ArrayList < Drawable > imageList) {
109.          super( ) ;
110.          this. imageList = imageList;
111.          this. ctx = ctx;
112.       }
113.       @Override
114.       public int getCount( ) {
115.          //TODO Auto – generated method stub
116.          return imageList. size( ) ;
117.       }
118.       @Override
119.       public Object getItem( int arg0) {
120.          //TODO Auto – generated method stub
121.          return this. imageList. get( arg0) ;
122.       }
123.       @Override
124.       public long getItemId( int arg0) {
125.          //TODO Auto – generated method stub
126.          return arg0;
127.       }
128.       @Override
129.       public View getView( int arg0, View arg1, ViewGroup arg2) {
130.          //TODO Auto – generated method stub
131.          ImageView image = new ImageView( ctx) ;
```

```
132.            image. setImageDrawable( this. imageList. get( arg0 ) ) ;
133.            image. setLayoutParams( new Gallery. LayoutParams(
134.            LinearLayout. LayoutParams. WRAP_CONTENT ,LinearLayout. LayoutParams. WRAP_CONTENT ) ) ;
135.            return image;
136.        }
137.    }
138.    @Override
139.    protected void onResume( ) {
140.        //TODO Auto – generated method stub
141.        super. onResume( ) ;
142.    }
143. }
```

5) 第五步

在包 szpt. android. ex03_07. listeners 下修改 AppListItemClickListener. java，将事件处理改为 Intent 跳转到 AppInfoActivity. java。

```
@Override
public void onItemClick( AdapterView < ? > arg0, View arg1, int arg2, long arg3) {
    Intent i = new Intent( ctx ,ApplyInfoActivity. class) ;
    HashMap < String, Object >  temp = listItem. get( arg2) ;
    int id = ( Integer) temp. get( " id" ) ;
    int icon = ( Integer) temp. get( " Icon" ) ;
    i. putExtra( " id" ,id) ;
    i. putExtra( " icon" , icon) ;
    tx. startActivity( i) ;
}
```

3. 代码分析

①第一步(3)点，在设计整个界面的布局的时候，注意使用布局容器的嵌套，在可视化的属性视图中进行设计时可以参考布局代码。

②第二步新增加的 AppInfoDetails 类，提供了 AppInfo 简要信息以外的更多 App 的信息，比如应用程序运行的预览图等。

③第三步修改了 ModelGenerator 类，模拟 App 数据的时候，新增加了一个分类和该分类下的 App 信息，并增加了一个方法返回该 App 的详细信息，这些返回的数据，都是提供给界面呈现用的。

④第四步：

- 第 35 ~ 37 行，获取 Intent 传递过来的参数，传递过来的是 App 的 id 和图标。
- 第 43 ~ 54 行，initViews( )方法，作用是从布局文件中获取需要进行设置的界面组件，存入成员变量中。

- 第 55～59 行，initAppInfoDetails( ) 方法，作用是获取 App 详细信息，初始化预览图片列表。
- 第 60～85 行，initInterface( ) 方法，作用是用获取的 Model 数据，初始化图形界面中的各个组件，特别是 Gallery apply_info_image 上的数据是使用自定义适配器对象，填充 ImageView 到 Gallery 上。

其中为了显示第几张图片显示在 Gallery 中的效果，在 Gallery 下面有个 LinearLayout apply_info_page 对象，该对象中含有多个 ImageView 对象，根据那张图片被显示，将其对应的 ImageView 中的设置成加亮的小圆点图片，其他都为灰色圆点图片。

- 第 109～149 行，ImageAdapter 作为 Gallery 的自定义适配器对象，其 getView( ) 方法返回的是一个 ImageView 对象，显示的就是 App 的预览图片。

⑤第五步修改 AppListItemClickListener 类，该类是单击分类中的某个具体 App 的事件处理器，方法中先获取选中 App 的 id 和 icon，放入 Intent 的 Bundle 中，随着 Intent 跳转到 AppInfoActivity 的同时，传递这两个数据。

任务4 集成应用商店

1. 任务说明

由于手机屏幕大小限制，使得手机难于像桌面程序一样可以同时显示大量信息，必须有一种途径，能够容纳多屏信息，同时又能够方便切换。使用 Tab 标签页控件，可以在同一个空间里放置更多内容。例如 Android 自带的拨号程序及通信录等，就使用了 Tab 标签功能。如图 3.15 所示，本项目就使用这两个控件，把前面开发的应用商店的典型功能通过 TabHost 和 TabWidget 进行了集成。

图 3.15　应用商城集成

2. 实现过程

1）第一步

如图 3.16 所示，新建总体布局文件 main_page.xml，TabHost 中包含线性布局，线性布局包含一个 FrameLayout 和一个 TabWidget 控件。

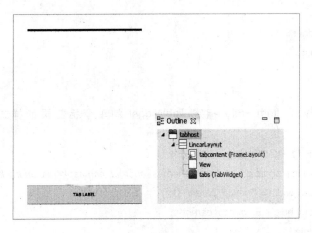

图 3.16 应用商城布局

```xml
<?xml version="1.0" encoding="utf-8"?>
<TabHost xmlns:android="http://schemas.android.com/apk/res/android"
    android:layout_width="match_parent"
    android:layout_height="match_parent"
    android:id="@android:id/tabhost"
    >
    <LinearLayout
        android:layout_width="fill_parent"
        android:layout_height="fill_parent"
        android:orientation="vertical"
        >
        <FrameLayout
            android:id="@android:id/tabcontent"
            android:layout_width="fill_parent"
            android:layout_height="fill_parent"
            android:layout_weight="1"
            >
        </FrameLayout>
        <View
            android:layout_width="fill_parent"
            android:layout_height="0.5dp"
            android:background="@android:color/darker_gray"
```

```xml
            />
            <TabWidget
                android:id="@android:id/tabs"
                android:layout_width="fill_parent"
                android:layout_height="wrap_content"
                >
            </TabWidget>
    </LinearLayout>
</TabHost>
```

2) 第二步

新建一个 tabhost_item.xml，指定 TabWidget 的单个标签页的样式，其中包含一个 TextView 文本框。

```xml
<?xml version="1.0" encoding="utf-8"?>
<LinearLayout xmlns:android="http://schemas.android.com/apk/res/android"
    android:layout_width="match_parent"
    android:layout_height="match_parent"
    android:orientation="vertical"
    android:background="@drawable/middle_item"
    >
    <TextView
        android:id="@+id/tabhost_item_tv"
        android:layout_width="wrap_content"
        android:layout_height="wrap_content"
        android:layout_gravity="center"
        android:textSize="15sp"
        android:text="TEXT"
        />
</LinearLayout>
```

3) 第三步

修改 MainActivity.java，加载布局文件，并设置相应的格式和事件处理。

1. **package** szpt.android.ex03_08;
2. **import** android.app.AlertDialog;
3. **import** android.app.Dialog;
4. **import** android.app.TabActivity;
5. **import** android.content.DialogInterface;
6. **import** android.content.Intent;
7. **import** android.os.Bundle;
8. **import** android.view.KeyEvent;
9. **import** android.view.LayoutInflater;
10. **import** android.view.View;

```
11.  import android.widget.TabHost;
12.  import android.widget.TabHost.OnTabChangeListener;
13.  import android.widget.TabWidget;
14.  import android.widget.TextView;
15.  public class MainActivity extends TabActivity {
16.      private TabWidget tabWidget;
17.      private TabHost tabHost;
18.      @Override
19.      protected void onCreate(Bundle savedInstanceState) {
20.          //TODO Auto-generated method stub
21.          super.onCreate(savedInstanceState);
22.          setContentView(R.layout.main_page);
23.          tabHost = (TabHost)findViewById(android.R.id.tabhost);
24.          LayoutInflater.from(this).inflate(R.layout.main_page,
                 tabHost.getTabContentView(),false);
25.          tabWidget = tabHost.getTabWidget();
26.          String[] title = {
27.              getResources().getString(R.string.feature),getResources().getString(R.string.hot,
                 getResources().getString(R.string.classify),getResources().getString
                 (R.string.system)
28.          };
29.          View[] viewTab = new View[title.length];
30.          for(int i=0;i<viewTab.length;i++){
31.              viewTab[i] = LayoutInflater.from(this).inflate(R.layout.tabhost_item,null);
32.              TextView tv = (TextView)viewTab[i].findViewById(R.id.tabhost_item_tv);
33.              tv.setText(title[i]);
34.          }
35.          tabHost.addTab(tabHost.newTabSpec("tab1")
36.              setIndicator(viewTab[0])
37.              setContent(new Intent(this,Feature.class)));
38.          tabHost.addTab(tabHost.newTabSpec("tab2")
39.              setIndicator(viewTab[1])
40.              setContent(new Intent(this,HotSoft.class)));
41.          tabHost.addTab(tabHost.newTabSpec("tab3")
42.              setIndicator(viewTab[2])
43.              setContent(new Intent(this,ClassifyActivity.class)));
44.          tabHost.addTab(tabHost.newTabSpec("tab4").setIndicator(viewTab[3])
45.              setContent(new Intent(this,ApplyListActivity.class)));
46.          View v = tabWidget.getChildAt(0);
47.          v.setBackgroundResource(R.drawable.middle_item_pressed);
```

```
48.         v.setClickable(false);
49.       tabHost.setOnTabChangedListener(new OnTabChangeListener() {
50.           @Override
51.           public void onTabChanged(String arg0) {
52.               //TODO Auto-generated method stub
53.               initTabWidget();
54.               int n = Integer.parseInt(" " + arg0.charAt(3));
55.               View v = tabWidget.getChildAt(n - 1);
56.               v.setBackgroundResource(R.drawable.middle_item_pressed);
57.               v.setClickable(false);
58.           }
59.       });
60.       tabHost.setCurrentTab(0);
61.   }
62.   private void initTabWidget() {
63.       for(int i = 0; i < tabWidget.getChildCount(); i++) {
64.           View v = tabWidget.getChildAt(i);
65.           v.setBackgroundResource(R.drawable.middle_item);
66.           v.setClickable(true);
67.       }
68.   }
69. }
```

3. 代码分析

①第一步。在这个布局文件中，需要特别注意的是包括一个 framelayout 和 tabwidget，framelayout 必须命名为 @android：id/tabcontent，tabwidget 必须命名为 @android：id/tabs。Activity 使用的是 TabActivity，其加载布局文件时，会根据布局文件中这两个控件的名字自动绑定这两个控件。

②第二步。这个布局文件是定义 TabHost 中标签页的显示方式，这里只是定义了标签页显示一个文本框，文本框所在的线性布局的背景色设置成了 android：background = " @drawable/middle_item"，这对应的是 middle_item.xml 文件，也就意味着线性布局根据所处的状态不同，背景色会不一样。

③第三步。
- 第 23~25 行，分别获取当前 TabActivity 中加载的 TabHost 对象和 TabWidget 对象。
- 第 26~34 行，使用资源文件获取标签文字和初始化 TabWidget 上要显示的各个 View 对象。
- 第 35~43 行，给当前的 TabHost 上加上 4 个 Tab 项，设置显示文字和对应的 Intent 跳转。
- 第 46~69 行，设置 tabHost 上的 TabChanged 事件监听器，当发生 TabChanged 事件的时

候，对 TabWidget 标签栏上的各个标签的状态进行调整。TabHost. OnTabChangeListener 接口中的 onTabChanged (String tabId)，当用户点击了不同的 Tab 选项卡时，触发该方法。

 功能拓展

如图 3.17 所示，标签页中"系统"标签页功能没有实现，请查找资料实现该功能，并且添加到当前工程中。

"系统"Activity 的功能是显示所有系统安装好的应用程序，并且在上面的搜索框中输入字母时能够实现动态过滤功能。

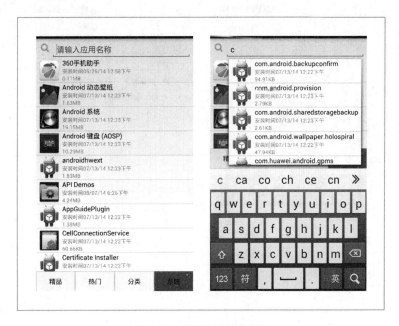

图 3.17　应用商城系统功能

提示：

①使用如下语句，获取系统安装好的应用列表

List < PackageInfo >　packageInfoList = his. getPackageManager ( ) . getInstalledPackages (0) ;

②使用上面的 Model 数据，就可以自定义适配器对象，把数据加载到一个 ListView 上。

③最上面的 AutoCompletTextView 也是一个适配器控件，加载数据后，能够显示多项内容。需要给它自定义一个适配器控件，并且实现带过滤功能。

 实战演练

利用所学高级 UI 控件实现图 3.18 所示掌上商城界面。

图 3.18 掌上商城

项目 4　Android 本地存储——掌上日记本

项目要点

- SQLite 本地存储实现。
- SimpleCursorAdapter 类的使用。

用户在使用应用程序过程中，需要将数据保存到本地手机上。一般使用文件、SQLite 将用户数据保存到本地。SQLite 是 Android 带的一个标准数据库，它支持 SQL 语句，是一个轻量级的嵌入式数据库。Android 应用软件的数据，包括文件、本地数据库数据，均为该应用软件所私有。为了提供应用程序之间的数据共享，提供了一种标准方式——内容提供器（content provider），供应用软件将私有数据开放给其他应用软件。

项目简介

本项目运用 Android 系统自带的 SQLite 数据库，做一个简易的日记本程序，实现对本地数据库的增、删、改、查操作，并且与 ListView 控件配合使用。

通过项目主要学习：

① 如何对 DatabaseHelper 和 SQLiteDatabase 封装，访问数据库。
② 如何利用 ContentValues 类来代替原始的 SQL 语句进行数据库的操作。
③ 如何使用 SimpleCursorAdapter 类和 ListView 配合进行 ListView 的显示。

效果如图 4.1 所示。

图 4.1　日记本

相关知识

SQLite 支持 NULL、INTEGER、REAL（浮点数字）、TEXT（字符串文本）和 BLOB（二

进制对象）5 种数据类型；SQLite 也接收 varchar（n）、char（n）、decimal（p，s）等数据类型，只不过在运算或保存时会转成对应的 5 种数据类型。

1. SQLiteDatabase

一个 SQLiteDatabase 的实例代表一个 SQLite 的数据库，通过 SQLiteDatabase 实例的一些方法，可以执行 SQL 语句，对数据库进行增、删、查、改操作。需要注意的是，数据库对于一个应用来说是私有的，并且在一个应用程序中，数据库的名字必须是唯一的。

1）SQLiteDatabase 基本操作

SQLite 可以解析大部分标准 SQL 语句。

（1）查询语句

select * from 表名 where 条件子句 group by 分组字句 having … order by 排序子句

例如：

select * from person

select * from person order by id desc

select name from person group by name having count(*)>1

分页 SQL 与 MySQL 类似，下面 SQL 语句获取 5 条记录，跳过前面 3 条记录：

select * from Account limit 5 offset 3 或者 select * from Account limit 3，5

（2）插入语句

insert into 表名（字段列表）values（值列表）

例如：

insert into person（name，age）values("小刀"，3）

（3）更新语句

update 表名 set 字段名＝值 where 条件子句

例如：

update person set name＝"飞刀" where id＝10

（4）删除语句

delete from 表名 where 条件子句

例如：

delete from person where id＝10

SQLiteDatabase 的类封装了一些操作数据库的 API，支持 CRUD 操作。

2）execSQL（）和 rawQuery（）方法

①execSQL（）：执行 insert、delete、update 和 CREATE TABLE 之类有更改行为的 SQL 语句；

②rawQuery（）方法可以执行 select 语句。

3）占位符参数（?）

execSQL（）方法的示例，如插入记录：

db.execSQL（"insert into person（name，age）values("小袁"，20)"）；

db.close（）；

SQLiteDatabase 类提供了一个重载后的 execSQL(String sql, Object[] bindArgs) 方法,使用这个方法可以解决前面提到的问题,因为这个方法支持使用占位符参数(?)。

使用例子如下:

db.execSQL("insert into person(name,age) values(?,?)", new Object[]{"斌哥", 25});

db.close();

execSQL(String sql, Object[] bindArgs) 方法的第一个参数为 SQL 语句,第二个参数为 SQL 语句中占位符参数的值,参数值在数组中的顺序要和占位符的位置对应。

2. SQLiteOpenHelper

这个类是一个辅助类,用来生成一个数据库,并对数据库的版本进行管理。

当在程序当中调用这个类的方法 getWritableDatabase(),或者 getReadableDatabase() 时,如果当时没有数据,那么 Android 系统会自动生成一个数据库。

SQLiteOpenHolper 是一个抽象类,通常需要继承它,并且实现里边的 3 个函数,具体函数如下所示。

①onCreate(SQLiteDatabase):在数据库第一次生成的时候会调用这个方法,一般在这个方法里边生成数据库表。

②onUpgrade(SQLiteDatabase, int, int):当数据库需要升级的时候,Android 系统会主动调用这个方法。一般在这个方法里边删除数据表,并建立新的数据表。是否需要做其他的操作完全取决于应用的需求。

③onOpen(SQLiteDatabase):这是当打开数据库时的回调函数,一般也不会用到。

3. Cursor

Cursor 在 Android 中是一个非常有用的接口,通过 Cursor 可以对从数据库查询出来的结果集进行随机的读写访问。

1) Cursor 注意事项

在理解和使用 Android Cursor 的时候,必须知道关于 Cursor 的几件事情:

①Cursor 是每行的集合。
②使用 moveToFirst() 定位第一行。
③必须知道每一列的名称。
④必须知道每一列的数据类型。
⑤Cursor 是一个随机的数据源。
⑥所有的数据都是通过下标取得。

当使用 SQLiteDatabase 的 query() 或 rawQuery() 方法时,就会得到 Cursor 对象,Cursor 所指向的就是每一条数据。Cursor 位于 android.database.Cursor 类,可见出它的设计是基于数据库服务产生的。

2) Cursor 常用函数

①moveToFirst()　　移动光标到第一行
②moveToLast()　　移动光标到最后一行

③moveToNext( )　　　移动光标到下一行
④isAfterLast( )　　　是否最后一行
⑤getColumnIndex(String columnName)　　返回指定列的名称,如果不存在返回 -1
⑥getString( )　　　返回指定列名称的值
⑦isBeforeFirst( )　　返回游标是否指向之前第一行的位置
⑧isAfterLast( )　　　返回游标是否指向第最后一行的位置

访问 Cursor 的下标可以获得其中的数据:
int nameColumn = cur. getColumnIndex(NAME);
String name = cur. getString(nameColumn);

再配上 for 循环,就取出我们需要的数据:
for(cur. moveToFirst( );! cur. isAfterLast( );cur. moveToNext( ))
{
　　int nameColumn = cur. getColumnIndex(NAME);
　　String name = cur. getString(nameColumn);
}

提示:在 Android 查询数据是通过 Cursor 类实现的。当使用 SQLiteDatabase. query( )方法时,就会得到 Cursor 对象,Cursor 所指向的就是每一条数据。

4. SQLite 基础操作

【例】 如图 4.2 所示 SQLite 基础操作实现以下功能:
①数据库的建立、删除。
②数据表的建立、删除。
③记录的添加、删除。
④使用 File Explore,查看数据库表。
⑤LogCat 查看打印日志。

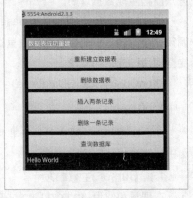

图 4.2　SQLite 基础操作

1) 实现过程
(1) 第一步
新建 main. xml,在线性布局中包含 5 个按钮。

1. < LinearLayout xmlns:android = " http://schemas. android. com/apk/res/android"
2. 　　android:orientation = " vertical" android:layout_width = " fill_parent"
3. 　　android:layout_height = " fill_parent">
4. 　< Button android:id = " @ + id/button1"
5. 　　　android:layout_width = " wrap_content"
6. 　　　android:layout_height = " wrap_content" android:text = " 重新建立数据表" />
7. 　< Button android:id = " @ + id/button2"
8. 　　　android:layout_width = " wrap_content"
9. 　　　android:layout_height = " wrap_content" android:text = " 删除数据表" />
10. 　< Button android:id = " @ + id/button3"
11. 　　　android:layout_width = " wrap_content"
12. 　　　android:layout_height = " wrap_content" android:text = " 插入两条记录" />

```
13.        <Button android:id = "@+id/button4"
14.            android:layout_width = "wrap_content"
15.            android:layout_height = "wrap_content" android:text = "删除一条记录"/>
16.        <Button android:id = "@+id/button5"
17.            android:layout_width = "wrap_content"
18.            android:layout_height = "wrap_content" android:text = "查询数据库"/>
19.    </LinearLayout>
```

（2）第二步

新建 ActivityMain.java，继承 Activity，在当前类中声明如下成员变量：

```
OnClickListener listener1 = null;
OnClickListener listener2 = null;
OnClickListener listener3 = null;
OnClickListener listener4 = null;
OnClickListener listener5 = null;
Button button1;
Button button2;
Button button3;
Button button4;
Button button5;
DatabaseHelper mOpenHelper;
private static final String DATABASE_NAME = "dbForTest.db";
private static final int DATABASE_VERSION = 1;
private static final String TABLE_NAME = "diary";
private static final String TITLE = "title";
private static final String BODY = "body";
```

（3）第三步

在 ActivityMain.java 中，新建一个内部静态类 DatabaseHelper 类，辅助数据库的建立。

```
private static class DatabaseHelper extends SQLiteOpenHelper {
    DatabaseHelper(Context context) {
        super(context, DATABASE_NAME, null, DATABASE_VERSION);
    }
    @Override
    public void onCreate(SQLiteDatabase db) {
        String sql = "CREATE TABLE " + TABLE_NAME + " (" + TITLE
            + " text not null, " + BODY + " text not null " + ");";
        Log.i("haiyang:createDB = ", sql);
        db.execSQL(sql);
    }
    @Override
    public void onUpgrade(SQLiteDatabase db, int oldVersion, int newVersion) {
```

            }
        }

(4) 第四步

新增加方法 prepareListener( ),使用该方法生成监听器。

```
private void prepareListener( ) {
        listener1 = new OnClickListener( ) {
            public void onClick( View v) {
                CreateTable( );
            }
        };
        listener2 = new OnClickListener( ) {
            public void onClick( View v) {
                dropTable( );
            }
        };
        listener3 = new OnClickListener( ) {
            public void onClick( View v) {
                insertItem( );
            }
        };
        listener4 = new OnClickListener( ) {
            public void onClick( View v) {
                deleteItem( );
            }
        };
        listener5 = new OnClickListener( ) {
            public void onClick( View v) {
                showItems( );
            }
        };
}
```

(5) 第五步

新增加方法 initLayout( ),为每个按钮设置监听器。

```
private void initLayout( ) {
        button1 = ( Button ) findViewById( R. id. button1 );
        button1. setOnClickListener( listener1 );
        button2 = Button ) findViewById( R. id. button2 );
        button2. setOnClickListener( listener2 );
        button3 = Button ) findViewById( R. id. button3 );
        button3. setOnClickListener( listener3 );
```

```java
        button4 = Button) findViewById( R. id. button4);
        button4. setOnClickListener( listener4);
        button5 = Button) findViewById( R. id. button5);
        button5. setOnClickListener( listener5);
    }
```

(6) 第六步

覆盖 onCreate( )方法。

```java
    super. onCreate( savedInstanceState);
    setContentView( R. layout. main);
    prepareListener( );
    initLayout( );
    mOpenHelper = new DatabaseHelper( this);
```

(7) 第七步

增加创建表、删除表、插入数据、删除数据和显示数据方法。

```java
    /*
     * 重新建立数据表
     */
    private void CreateTable( ) {
        SQLiteDatabase db = mOpenHelper. getWritableDatabase( );
        String sql = " CREATE TABLE " + TABLE_NAME + " ( " + TITLE
                + " text not null, " + BODY + " text not null " + " );";
        Log. i( " haiyang:createDB = ", sql);
        try {
            db. execSQL( " DROP TABLE IF EXISTS diary");
            db. execSQL( sql);
            setTitle( " 数据表成功重建");
        } catch ( SQLException e) {
            setTitle( " 数据表重建错误");
        }
    }
    /*
     * 删除数据表
     */
    private void dropTable( ) {
        SQLiteDatabase db = mOpenHelper. getWritableDatabase( );
        String sql = " drop table " + TABLE_NAME;
        try {
            db. execSQL( sql);
            setTitle( " 数据表成功删除:" + sql);
```

```java
        } catch(SQLException e) {
            setTitle("数据表删除错误");
        }
    }
    /*
     * 插入两条数据
     */
    private void insertItem() {
        SQLiteDatabase db = mOpenHelper.getWritableDatabase();
        String sql1 = "insert into " + TABLE_NAME + " (" + TITLE + ", " + BODY
                + ") values('haiyang', 'android 的发展真是迅速啊');";
        String sql2 = "insert into " + TABLE_NAME + " (" + TITLE + ", " + BODY
                + ") values('icesky', 'android 的发展真是迅速啊');";
        try {
            Log.i("haiyang:sql1 = ", sql1);
            Log.i("haiyang:sql2 = ", sql2);
            db.execSQL(sql1);
            db.execSQL(sql2);
            setTitle("插入两条数据成功");
        } catch(SQLException e) {
            setTitle("插入两条数据失败");
        }
    }
    /*
     * 删除其中的一条数据
     */
    private void deleteItem() {
        try {
            SQLiteDatabase db = mOpenHelper.getWritableDatabase();
            db.delete(TABLE_NAME, "title = 'haiyang'", null);
            setTitle("删除 title 为 haiyang 的一条记录");
        } catch(SQLException e) {
            setTitle("数据表删除其中的一条数据时错误");
        }
    }
    /*
     * 在屏幕的 Title 区域显示当前数据表当中的数据的条数。
     */
    private void showItems() {
        try {
```

```
SQLiteDatabase db = mOpenHelper.getReadableDatabase();
String col[] = {TITLE, BODY};
Cursor cur = db.query(TABLE_NAME, col, null, null, null, null, null);
Integer num = cur.getCount();
setTitle(Integer.toString(num) + "条记录");
}catch(SQLException e){
    setTitle("数据表显示错误");
}
```

2）代码分析

①SQLiteDatabase db = mOpenHelper.getReadableDatabase();首先得到一个可写的数据库。

②Cursor cur = db.query(TABLE_NAME, col, null, null, null, null, null);将查询到的数据放到一个 Cursor 当中。这个 Cursor 里边封装了这个数据表 TABLE_NAME 当中的所有条列。query()方法相当的有用，下面简单介绍。

- 第一个参数是数据库里边表的名字，比如本例中表的名字就是 TABLE_NAME，也就是" diary"。
- 第二个字段是想要返回数据包含的列的信息。本例中想要得到的列有 title、body。把这两个列的名字放到字符串数组里边来。
- 第三个参数为 selection，相当于 SQL 语句的 where 部分，如果想返回所有的数据，那么就直接置为 null。
- 第四个参数为 selectionArgs。在 selection 部分，有可能用到"?"，那么在 selectionArgs 定义的字符串会代替 selection 中的"?"。
- 第五个参数为 groupBy。定义查询出来的数据是否分组，如果为 null 则说明不用分组。
- 第六个参数为 having，相当于 SQL 语句当中的 having 部分。
- 第七个参数为 orderBy，来描述期望的返回值是否需要排序，如果设置为 null 则说明不需要排序。

③Integer num = cur.getCount();通过 getCount()方法，可以得到 cursor 当中数据的个数。

任务1 搭建布局文件

1. 任务说明

我们首先搭建日记本程序的布局文件，主要包括主界面、新增日记界面和日记列表界面。

2. 实现过程

1）第一步

搭建主界面 diary_list.xml。

<?xml version="1.0" encoding="utf-8"?>

```
<LinearLayout xmlns:android = " http://schemas.android.com/apk/res/android"
    android:layout_width = " fill_parent"
    android:layout_height = " wrap_content">
    <ListView
        android:id = " @ + id/android:list"
        android:layout_width = " fill_parent"
        android:layout_height = " wrap_content" / >
    <TextView
        android:id = " @ + id/android:empty"
        android:layout_width = " fill_parent"
        android:layout_height = " wrap_content"
        android:text = " 您还没有开始写日记呢！点击下边的 Menu 按钮开始写日记吧：）" / >
</LinearLayout>
```

注意：

①当 ListView 不为空时，只显示 listview；反之显示 TextView。这和 java 文件里的代码没有关系，是由 TextView 的 **android：id = "@ + id/android：empty**" 属性实现的。

②TextView 的 android：id = " @ + id/android：empty 属性决定当 ListView 为空时自动显示 TextView。忽略布局里的 ListView 的存在，认为整个布局里只有 TextView。

2）第二步

如图 4.3 所示，新建日记添加界面 diary_edit.xml。

图 4.3　添加日记界面

```
//LinearLayout 的嵌套布局
<? xml version = " 1.0" encoding = " utf - 8" ? >
<LinearLayout xmlns:android = " http://schemas.android.com/apk/res/android"
    android:layout_width = " fill_parent"
    android:layout_height = " fill_parent"
    android:orientation = " vertical">
    <LinearLayout
        android:layout_width = " fill_parent"
        android:layout_height = " wrap_content"
        android:orientation = " horizontal">
        <TextView
            android:layout_width = " wrap_content"
            android:layout_height = " wrap_content"
            android:text = " @string/title" / >
        <EditText
            android:id = " @ + id/title"
```

```
            android:layout_width = "wrap_content"
            android:layout_height = "wrap_content"
            android:layout_weight = "1" />
    </LinearLayout>
    <TextView
        android:layout_width = "wrap_content"
        android:layout_height = "wrap_content"
        android:text = "@string/body" />
    <EditText
        android:id = "@+id/body"
        android:layout_width = "fill_parent"
        android:layout_height = "wrap_content"
        android:layout_weight = "1"
        android:gravity = "top"
        android:scrollbars = "vertical" />
    <Button
        android:id = "@+id/confirm"
        android:layout_width = "wrap_content"
        android:layout_height = "wrap_content"
        android:text = "@string/confirm" />
</LinearLayout>
```

3) 第三步

如图 4.4 所示，新建列表单项格式 diary_row.xml。

图 4.4 日记列表

```
<?xml version = "1.0" encoding = "utf-8"?>
<RelativeLayout xmlns:android = "http://schemas.android.com/apk/res/android"
    android:id = "@+id/row"
    android:layout_width = "fill_parent"
    android:layout_height = "fill_parent">
<TextView
    android:id = "@+id/text1"
    android:layout_width = "wrap_content"
    android:layout_height = "35px"
    android:layout_marginTop = "10dip"
    android:maxWidth = "200dip"
    android:text = "第一组第一项"
    android:textSize = "16sp" />
<TextView
    android:id = "@+id/created"
    android:layout_width = "wrap_content"
```

```
            android:layout_height = "35px"
            android:layout_alignParentRight = "true"
            android:layout_marginLeft = "10dip"
            android:layout_marginTop = "10dip"
            android:text = "2013年11月11号"/>
</RelativeLayout>
```

4）第四步

在 strings.xml 中，增加项目所需常量。

```
<?xml version = "1.0" encoding = "utf-8"?>
<resources>
    <string name = "app_name">个人日记本</string>
    <string name = "menu_insert">添加一篇新日记</string>
    <string name = "menu_delete">删除一篇日记</string>
    <string name = "title">标题</string>
    <string name = "body">内容</string>
    <string name = "confirm">确定</string>
    <string name = "edit_diary">编辑日记</string>
</resources>
```

3. 代码分析

在第一个布局文件中，特别注意：

①当 ListView 不为空时，只显示 listview；反之显示 TextView。这和 java 文件中的代码没有关系，是由 TextView 的 android:id = "@ +id/android:empty 属性实现的。

②TextView 的 android:id = "@ +id/android:empty 属性决定当 ListView 为空时自动显示 TextView。忽略布局里的 ListView 的存在，认为整个布局里只有 TextView。

任务2 封装数据操作——适配器

1. 任务说明

类 DiaryDbAdapter 辅助数据库的操作。这个类封装了 DatabaseHelper 和 SQLiteDatabase 类的操作，使得对数据库的操作更加安全和方便。

2. 实现过程

1. **import** java.util.Calendar;
2. **import** android.content.ContentValues;
3. **import** android.content.Context;
4. **import** android.database.Cursor;
5. **import** android.database.SQLException;
6. **import** android.database.sqlite.SQLiteDatabase;
7. **import** android.database.sqlite.SQLiteOpenHelper;

8. **public class** DiaryDbAdapter {
9.   **public static final String** KEY_TITLE = "title";
10.   **public static final String** KEY_BODY = "body";
11.   **public static final String** KEY_ROWID = "_id";
12.   **public static final String** KEY_CREATED = "created";
13. //日记撰写时间
14.   **private static final** String TAG = "DiaryDbAdapter";
15.   **private** DatabaseHelper mDbHelper;
16.   **private** SQLiteDatabase mDb;
17.   **private static final** String DATABASE_CREATE = "create table diary(_id integer primary key autoincrement, "
18.   + "title text not null, body text not null, created text not null);";
19.   **private static final** String DATABASE_NAME = "database";
20.   **private static final** String DATABASE_TABLE = "diary";
21.   **private static final int** DATABASE_VERSION = 1;
22.   **private final** Context mCtx; // final 的定义是申请一片常量池，其值不可改变
23. //内部静态类 DatabaseHelper
24.   **private static class** DatabaseHelper **extends** SQLiteOpenHelper {
25.   DatabaseHelper(Context context) {
26.   **super**(context, DATABASE_NAME, **null**, DATABASE_VERSION);
27.   }
28.   @Override
29.   **public void** onCreate(SQLiteDatabase db) {
30.   db.execSQL(DATABASE_CREATE);
31.   }
32.   @Override
33.   **public void** onUpgrade(SQLiteDatabase db, **int** oldVersion, **int** newVersion) {
34.   db.execSQL("DROP TABLE IF EXISTS diary");
35.   onCreate(db);
36.   }
37.   }
38. //构造器
39. // Context 字面意思是上下文，在 Android 中可以进行很多操作，最主要的功能是加载和访问资源。
     在 Android 中有两种 Context：一种是 application context，一种是 activity context。
40.   **public** DiaryDbAdapter(Context ctx) {
41.   **this**.mCtx = ctx;
42.   }
43. // DiaryDbAdapter
44.   **public** DiaryDbAdapter open() **throws** SQLException {

```
45.     mDbHelper = new DatabaseHelper(mCtx);
46.     mDb = mDbHelper.getWritableDatabase();
47.     return this;
48.   }
49.   public void closeclose(){
50.     mDbHelper.close();
51.   }
52. // long l = 9223372036854775806L, LONG 最大值是 9223372036854775807
53.   public long createDiary(String title, String body) {
54.     ContentValues initialValues = new ContentValues();
55.     initialValues.put(KEY_TITLE, title);
56.     initialValues.put(KEY_BODY, body);
57.     Calendar calendar = Calendar.getInstance();
58.     String created = calendar.get(Calendar.YEAR) + "年"
59.         + calendar.get(Calendar.MONTH) + "月"
60.         + calendar.get(Calendar.DAY_OF_MONTH) + "日"
61.         + calendar.get(Calendar.HOUR_OF_DAY) + "时"
62.         + calendar.get(Calendar.MINUTE) + "分";
63.     initialValues.put(KEY_CREATED, created);
64.     return mDb.insert(DATABASE_TABLE, null, initialValues);
65.   }
66.   public boolean deleteDiary(long rowId) {
67.     return mDb.delete(DATABASE_TABLE, KEY_ROWID + "=" + rowId, null) > 0;
68.   }
69.   public Cursor getAllNotes() {
70.     return mDb.query(DATABASE_TABLE, new String[] { KEY_ROWID, KEY_TITLE,
71.         KEY_BODY, KEY_CREATED }, null, null, null, null, null);
72.   }
73.   public Cursor getDiary(long rowId) throws SQLException {
74.     Cursor mCursor =
75.       mDb.query(true, DATABASE_TABLE, new String[] { KEY_ROWID, KEY_TITLE,
76.           KEY_BODY, KEY_CREATED }, KEY_ROWID + "=" + rowId, null, null,
77.           null, null, null);
78.     if (mCursor != null) {
79.         mCursor.moveToFirst();
80.     }
81.     return mCursor;
82.   }
83.   public boolean updateDiary(long rowId, String title, String body) {
84.     ContentValues args = new ContentValues();
```

```
85.   /* ContentValues 类和 Hashtable 类似,它也是负责存储一些名值对,但是它存储的名值对当中的
        名是一个 String 类型,而值都是基本类型。在这里将要插入的值都放到一个 ContentValues 的实
        例当中,然后执行插入操作 */
86.   args. put( KEY_TITLE, title );
87.   args. put( KEY_BODY, body );
88.   Calendar calendar = Calendar. getInstance( );
89.   String created = calendar. get( Calendar. YEAR) + "年"
90.            + calendar. get( Calendar. MONTH) + "月"
91.            + calendar. get( Calendar. DAY_OF_MONTH) + "日"
92.            + calendar. get( Calendar. HOUR_OF_DAY) + "时"
93.            + calendar. get( Calendar. MINUTE) + "分";
94.   args. put( KEY_CREATED, created );
95.   return mDb. update( DATABASE_TABLE, args, KEY_ROWID + " = " + rowId, null) > 0;
96.   }
97.  }
```

3. 代码分析

①在 DiaryDbAdapter 类里,针对日记的增、删、查、改,为调用者提供如下接口方法,封装数据库的操作细节。

- open( ):调用这个方法后,如果数据库还没有建立,那么会建立数据库;如果数据库已经建立了,那么会返回可写的数据库实例。
- close( ):调用此方法,DatabaseHelper 会关闭对数据库的访问。
- createDiary(String title, String body):通过一个 title 和 body 字段在数据库当中创建一条新的记录。
- deleteDiary(long rowId):通过记录的 id,删除数据库中指定的记录。
- getAllNotes( ):得到 diary 表中所有的记录,并且以一个 Cursor 的形式进行返回。
- getDiary(long rowId):通过记录的主键 id,得到特定的一条记录。
- updateDiary(long rowId, String title, String body):更新主键 id 为 rowId 那条记录中的两个字段 title 和 body 字段的内容。

②ContentValues 类。ContentValues 类和 Hashtable 类似,它也是负责存储一些键值对,存储的键值对当中的键是一个 String 类型,而值都是基本类型。通过 SQL 语句进行插入操作。SQL 语句的好处是比较直观,但是容易出错。在这个例子当中有更好的办法,将要插入的值都放到一个 ContentValues 的实例当中,然后执行插入操作,具体代码如下所示:

```
public long createDiary( String title, String body) {
    ContentValues initialValues = new ContentValues( );
    initialValues. put( KEY_TITLE, title );
    initialValues. put( KEY_BODY, body );
    Calendar calendar = Calendar. getInstance( );
    //生成年月日字符串
```

```
            String created = calendar.get(Calendar.YEAR) + "年" + calendar.get(Calendar.MONTH) + "
月" + calendar.get(Calendar.DAY_OF_MONTH) + "日" + calendar.get(Calendar.HOUR_OF_DAY) + "
时" + calendar.get(Calendar.MINUTE) + "分";
            initialValues.put(KEY_CREATED, created);
            return mDb.insert(DATABASE_TABLE, null, initialValues);
        }
            ContentValues initialValues = new ContentValues();     //实例化一个contentValues 类
            initialValues.put (KEY_TITLE, title); //将列名和对应的列值放置到initialValues 里边
            mDb.insert(DATABASE_TABLE, null, initialValues);    //负责插入一条新的纪录,如果插入成功
            //则会返回这条记录的id,如果插入失败会返回-1
```

在更新一条记录的时候,也是采用ContentValues 的这套机制,具体代码如下所示:

```
public boolean updateDiary(long rowId, String title, String body){
        ContentValues args = new ContentValues();
        args.put(KEY_TITLE, title);
        args.put(KEY_BODY, body);
        Calendar calendar = Calendar.getInstance();
        String created = calendar.get(Calendar.YEAR) + "年"
            + calendar.get(Calendar.MONTH) + "月"
            + calendar.get(Calendar.DAY_OF_MONTH) + "日"
            + calendar.get(Calendar.HOUR_OF_DAY) + "时"
            + calendar.get(Calendar.MINUTE) + "分";
        args.put(KEY_CREATED, created);
        return mDb.update(DATABASE_TABLE, args, KEY_ROWID + " = " + rowId, null) > 0;
}
```

任务3 搭建主程序

1. 任务说明

构建日记本的主程序,并且添加相应的事件处理。

2. 实现过程

① 修改项目主程序ActivityMain.java,增加成员变量。

修改主程序ActivityMain.java 为一个ListActivity 文件:

　　public class ActivityMain extends ListActivity {...}

② 在onCreate()方法导入布局文件。

　　super.onCreate(savedInstanceState);
　　setContentView(R.layout.*diary_list*);

③ 增加菜单按钮,并添加事件处理。成员变量:

　　private static final int INSERT_ID = Menu.*FIRST*;

```java
private static final int DELETE_ID = Menu.FIRST + 1;
```
④实现onCreateOptionsMenu()和onMenuItemSelected()两个方法。
```java
@Override
public boolean onCreateOptionsMenu(Menu menu) {
    super.onCreateOptionsMenu(menu);
    menu.add(0, INSERT_ID, 0, R.string.menu_insert);
    menu.add(0, DELETE_ID, 0, R.string.menu_delete);
    return true;
}

@Override
public boolean onMenuItemSelected(int featureId, MenuItem item) {
    switch(item.getItemId()) {
    case INSERT_ID:
        //跳转到添加日志界面
        createDiary();
        return true;
    case DELETE_ID:
        //删除选中日志
        mDbHelper.deleteDiary(getListView().getSelectedItemId());
        renderListView();
        return true;
    }
    return super.onMenuItemSelected(featureId, item);
}
```
⑤增加成员变量。
```java
private static final int ACTIVITY_CREATE = 0;
private static final int ACTIVITY_EDIT = 1;
private DiaryDbAdapter mDbHelper;
private Cursor mDiaryCursor;
```
⑥添加renderListView()方法，主要作用是取出所有记录，更新列表适配器，Activity.startManagingCursor()方法：将获得的Cursor对象交与Activity管理，这样Cursor对象的生命周期便能与当前的Activity自动同步，省去了自己管理Cursor。
```java
private void renderListView() {
    mDiaryCursor = mDbHelper.getAllNotes();
    startManagingCursor(mDiaryCursor);
    String[] from = new String[] { DiaryDbAdapter.KEY_TITLE,
            DiaryDbAdapter.KEY_CREATED };
    int[] to = new int[] { R.id.text1, R.id.created };
    SimpleCursorAdapter notes = new SimpleCursorAdapter(this,
            R.layout.diary_row, mDiaryCursor, from, to);
```

```
        setListAdapter(notes);
    }
```
⑦增加 createDiary()方法，跳转页面。
```
    private void createDiary() {
        Intent i = new Intent(this, ActivityDiaryEdit.class);
        startActivityForResult(i, ACTIVITY_CREATE);
    }
```
Intent 应该算是 Android 中特有的。可以在 Intent 中指定程序要执行的动作（如 view、edit、dial），以及程序执行到该动作时所需要的资料。都指定好后，只要调用 startActivity()，Android 系统会自动寻找最符合指定要求的应用程序，并执行该程序。

⑧使用回调方法，更新列表框。
```
    @Override
    protected void onActivityResult(int requestCode, int resultCode,
            Intent intent) {
        super.onActivityResult(requestCode, resultCode, intent);
        renderListView();
    }
```
startActivityForResult 的主要作用是它可以回传数据，假设有两个页面，首先进入第一个页面，里面有一个按钮，用于进入下一个页面，当进入下一个页面时，进行设置操作，并在其 finish 动作或者 back 动作后，将设置的值回传给第一个页面，从而第一个页面来显示所得到的值。

⑨点击的事件处理代码。需要对 position 和 id 进行区分。position 指的是点击 ViewItem 在当前 ListView 中的位置，每一个和 ViewItem 绑定的数据都有一个 id，通过这个 id 可以找到那条数据。
```
    @Override
    protected void onListItemClick(ListView l, View v, int position, long id) {
        super.onListItemClick(l, v, position, id);
        Cursor c = mDiaryCursor;
        c.moveToPosition(position);
        Intent i = new Intent(this, ActivityDiaryEdit.class);
        i.putExtra(DiaryDbAdapter.KEY_ROWID, id);
        i.putExtra(DiaryDbAdapter.KEY_TITLE, c.getString(c
                .getColumnIndexOrThrow(DiaryDbAdapter.KEY_TITLE)));
        i.putExtra(DiaryDbAdapter.KEY_BODY, c.getString(c
                .getColumnIndexOrThrow(DiaryDbAdapter.KEY_BODY)));
        startActivityForResult(i, ACTIVITY_EDIT);
    }
```
⑩在 OnCreate()方法中增添三条语句，进行初始化的准备工作
```
        mDbHelper = new DiaryDbAdapter(this);
```

```
mDbHelper.open();
renderListView();
```

任务4　编写日记功能

1. 任务说明

新建新的界面，实现日记的编辑。

2. 实现过程

```
import android.app.Activity;
import android.content.Intent;
import android.os.Bundle;
import android.view.View;
import android.widget.Button;
import android.widget.EditText;
public class ActivityDiaryEdit extends Activity {
    private EditText mTitleText;
    private EditText mBodyText;
    private Long mRowId;
    private DiaryDbAdapter mDbHelper;
    @Override
    protected void onCreate(Bundle savedInstanceState) {
        super.onCreate(savedInstanceState);
        mDbHelper = new DiaryDbAdapter(this);
        mDbHelper.open();
        setContentView(R.layout.diary_edit);
        mTitleText = (EditText) findViewById(R.id.title);
        mBodyText = (EditText) findViewById(R.id.body);
        Button confirmButton = (Button) findViewById(R.id.confirm);
        mRowId = null;
        // 每一个 intent 都会带一个 Bundle 型的 extras 数据
        Bundle extras = getIntent().getExtras();
        if(extras != null) {
            String title = extras.getString(DiaryDbAdapter.KEY_TITLE);
            String body = extras.getString(DiaryDbAdapter.KEY_BODY);
            mRowId = extras.getLong(DiaryDbAdapter.KEY_ROWID);
            if(title != null) {
                mTitleText.setText(title);
            }
            if(body != null) {
```

```
            mBodyText.setText(body);
        }
    }
    confirmButton.setOnClickListener(new View.OnClickListener() {
        public void onClick(View view) {
            String title = mTitleText.getText().toString();
            String body = mBodyText.getText().toString();
            if(mRowId != null) {
                mDbHelper.updateDiary(mRowId, title, body);
            } else
                mDbHelper.createDiary(title, body);
            Intent mIntent = new Intent();
            setResult(RESULT_OK, mIntent);
            finish();
        }
    });
}
```

在文件 AndroidManifest.xml 增加 ActivityDiaryEdit.java 的注册信息：

<activity android:name=".ActivityDiaryEdit"/>

注意：添加的信息位于 <application>...</application> 中。

功能拓展

查资料学习 ContentProvider 共享机制，修改掌上日记本程序，把本地数据库中存放的日记数据，通过 ContentProvider 共享机制，提供开放的接口。

实战演练

完成个人记账本。请大家相互调研个人使用手机进行记账的需求，并对比 AppStore 中已有 App 的优劣，利用高级 UI 和 SQLite 本地数据库技术完成一个"个人记账本 App"。

项目 5 Android 网络通信——天气预报

 项目要点

- 了解 Android 网络访问基本概念。
- 熟练网络数据解析的方式。
- 多线程在网络通信中的应用。

前面学习了各种 UI 组件、本地存储，了解 Activity、Intent 和 SQLite 知识，但是都只是涉及手机本地信息。随着移动互联 4G 的普及，一个应用如果缺少网络的支持，将缺乏生命力。本项目将学习 Android 的网络通信编程。

Android 的网络通信编程主要涉及：①如何访问外网的数据，数据的来源可以是应用服务器、Web 服务器上的某种应用服务；②如何解析获取到的数据，一般数据的格式是字节流、XML 格式化数据或者 json 数据等，获取到数据后必须解析出需要的数据，加载在 UI 上，显示给用户；③如何将解析出的数据显示在 Android 的界面中，要合理显示在 Android 界面中，涉及多线程技术的应用，在 Android 里面多线程机制与 J2SE 中存在一定的差异，在多线程一节中会有详细介绍。

 项目简介

本项目实现图 5.1 所示的界面。运行程序后，点击右边的按钮，屏幕显示选择城市的天气状况。

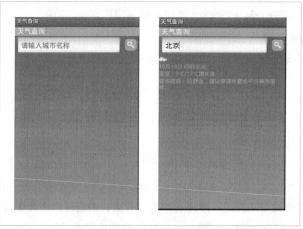

图 5.1 天气预报系统

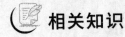 相关知识

1. Android 网络通信简介

如图 5.2 所示，Android 网络通信沿用了标准 Java 接口 java.net.* 中的类，用于网络通信，还提供了一些额外的类：①Apache 接口 org.apache.http.*；②Android 网络接口 android.net.*，用于扩充功能和特定于 Android 的一些网络功能。

包	描述
java.net	提供与联网有关的类，包括流和数据包（datagram）sockets、Internet 协议和常见 HTTP 处理。该包是一个多功能网络资源。有经验的 Java 开发人员可以立即使用这个熟悉的包创建应用程序。
java.io	虽然没有提供显式的联网功能，但是仍然非常重要。该包中的类由其他 Java 包中提供的 socket 和连接使用。它们还用于与本地文件（在与网络进行交互时会经常出现）的交互。
java.nio	包含表示特定数据类型的缓冲区的类。适合用于两个基于 Java 语言的端点之间的通信。
org.apache.*	表示许多为 HTTP 通信提供精确控制和功能的包。可以将 Apache 视为流行的开源 Web 服务器。
android.net	除核心 java.net.* 类以外，包含额外的网络访问 socket。该包包括 URI 类，后者频繁用于 Android 应用程序开发，而不仅仅是传统的联网方面
android.net.http	包含处理 SSL 证书的类。
android.net.wifi	包含在 Android 平台上管理有关 WiFi（802.11 无线 Ethernet）所有方面的类。并不是所有设备都配备了 WiFi 功能，特别是 Android 在 Motorola 和 LG 等手机制造商的"翻盖手机"领域获得了成功。
android.telephony.gsm	包含用于管理和发送 SMS（文本）消息的类。一段时间后，可能会引入额外的包来为非 GSM 网络提供类似的功能，比如 CDMA 或 android.telephony.cdma 等网络。

图 5.2 网络包结构

Android 的网络编程分为基于 Socket 的和基于 HTTP 协议的两种。

1）Socket 编程

基于 Socket 的用法类似于 J2SE 中的 client-server 编程方式。

（1）服务器端

①先启动一个服务器端的 socket：ServerSocket svr = new ServerSocket(8989)；

②开始侦听请求：Socket s = svr.accept()；

③取得输入和输出：

DataInputStream dis = new DataInputStream(s.getInputStream())；

DataOutputStream dos = new DataOutputStream(s.getOutputStream())；

④Socket 的交互通过流来完成，即传送的是字节流，因此任何文件都可以在上面传送。

用 DataInputStream/DataOutputStream 来进行包装是因为需要它们对基本数据类型的读写功能，如 readInt()、writeInt()、readUTF()、writeUTF() 等。

(2) 客户端

①发起一个 Socket 连接:

Socket s = new Socket("192.168.1.200",8989);

②取得输入和输出:

DataInputStream dis = new DataInputStream(s.getInputStream());

DataOutputStream dos = new DataOutputStream(s.getOutputStream());

2) HTTP 协议

万维网上的 Web 程序,遵循的都是 HTTP 协议,请求相应协议,在该种编程模式下,使用 Android 客户端模拟 HTTP 协议,与 Web Server 进行交互,java.net.*提供了与联网有关的类。

(1) 简单流程

①创建 URL。

②创建 URLConnection/HttpURLConnection 对象。

③设置连接参数、连接到服务器、向服务器写数据。

④从服务器读取数据。

(2) 具体步骤

①一般是发送请求到某个应用服务器。此时需要用到 HttpURLConnection。

HttpURLConnection urlConn = new URL("http://www.google.com").openConnection();

②设置标志:

urlConn.setDoOutput(true); urlConn.setDoInput(true);

//post 的情况下需要设置 DoOutput 为 true

urlConn.setRequestMethod("POST");

urlConn.setUseCache(false);//设置是否用缓存

urlConn.setRequestProperty("Content-type","application/x-www-form-urlencoded");

//设置 content-type 获得输出流,便于向服务器发送信息

③DataOutputStream dos = new DataOutputStream(urlConn.getOutputStream());

向流里面写请求参数:

dos.writeBytes("name = " + URLEncoder.encode("chenmouren","gb2312"));

dos.flush();

dos.close();//发送完后马上关闭

④获得输入流,取数据。

BufferReader reader = new BufferedReader(new InputStreamReader(urlConn.getInputStream()));

reader.readLine();//用! = null 判断是否结束

reader.close();

⑤读完后关闭 connection。

urlConn.disconnect();

2. 数据的解析——XML 文件的解析

一般从网络获取的数据主要以 3 种形式存在：①字节流；②XML 数据；③JSON 格式。其中，第一种需要自己解析字节流，根据协议不同，解析方式不同；第二种用得比较广泛，用于表示结构化的数据，解析方式相对固定，便于理解；第三种是很多 Web 服务提供的返回数据格式，也是一种简化的结构化数据，也有固定解析方式。在此主要针对 XML 数据给出 3 种不同解析方式，然后比较它们之间的优缺点。

1）XML 数据解析——SAX

SAX 是一个解析速度快并且占用内存少的 XML 解析器，非常适合用于 Android 等移动设备。SAX 解析 XML 文件采用的是事件驱动，也就是说，它并不需要解析完整个文档。在按内容顺序解析文档的过程中，SAX 会判断当前读到的字符是否是合法 XML 语法中的某部分，如果符合就会触发事件。所谓事件，其实就是一些回调（callback）方法，这些方法（事件）定义在 ContentHandler 接口。下面是一些 ContentHandler 接口常用的方法：

①startDocument( )

当遇到文档的开头的时候，调用这个方法，可以在其中做一些预处理的工作。

②endDocument( )

和上面的方法相对应，当文档结束的时候，调用这个方法，可以在其中做一些善后的工作。

③startElement(String namespaceURI, String localName, String qName, Attributes atts)

当读到一个开始标签的时候，会触发这个方法。namespaceURI 就是命名空间，localName 是不带命名空间前缀的标签名，qName 是带命名空间前缀的标签名。通过 atts 可以得到所有的属性名和相应的值。要注意的是，SAX 一个重要的特点就是它的流式处理，当遇到一个标签的时候，它并不会记录下以前所碰到的标签，也就是说，在 startElement( ) 方法中，所知道的信息只有标签的名字和属性，标签的嵌套结构、上层标签的名字、是否有子元素等其他与结构相关的信息都是不得而知的，都需要程序来完成。这使得 SAX 在编程处理上没有 DOM 方便。

④endElement(String uri, String localName, String name)

这个方法和上面的方法相对应，在遇到结束标签的时候，调用这个方法。

⑤characters(char[ ] ch, int start, int length)

这个方法用来处理在 XML 文件中读到的内容，第一个参数用于存放文件的内容，后面两个参数是读到的字符串在这个数组中的起始位置和长度，使用 new String(ch,start,length) 可以获取内容。

只要为 SAX 提供实现 ContentHandler 接口的类，那么该类就可以得到通知事件（实际上就是 SAX 调用了该类中的回调方法）。因为 ContentHandler 是一个接口，在使用的时候可能会有些不方便，因此，SAX 还为其指定了一个 Helper 类：DefaultHandler，它实现了这个接口，但是其所有的方法体都为空。在实现的时候，只需要继承这个类，然后重载相应的方法即可。使用 SAX 解析 itcast.xml 的代码如下：

```
public static List < Person > readXML( InputStream inStream ) {
    try {
        SAXParserFactory spf = SAXParserFactory. newInstance( );
```

```
            SAXParser saxParser = spf.newSAXParser();  //创建解析器
            //设置解析器的相关特性,http://xml.org/sax/features/namespaces = true 表示开启命名空间特性
            saxParser.setProperty("http://xml.org/sax/features/namespaces",true);
            XMLContentHandler handler = new XMLContentHandler();
            saxParser.parse(inStream,handler);
            inStream.close();
            return handler.getPersons();
        } catch (Exception e) {
            e.printStackTrace();
        }
    return null;
}
```

SAX 支持已内置到 JDK1.5 中,用户无须添加任何的 jar 文件。

【例 5-1】 通过 SAX,实现将 SD 卡上的 person.xml 文件中的两个 person 信息解析出来,显示在 ListView 中,如图 5.3 所示。

新建一个文件 person.xml,并 push 到手机上的 SD 卡。

```xml
<?xml version="1.0" encoding="UTF-8"?>
<persons>
    <person id="23">
        <name>李明</name>
        <age>30</age>
    </person>
    <person id="20">
        <name>李向梅</name>
        <age>25</age>
    </person>
</persons>
```

图 5.3　学生列表

(1) 实现过程

参照项目 1,在 Eclipse 中创建名为 ex05_1SAX 的工程。

① 编写 XML 布局文件,在 res/layout 目录下编写 ex02.xml,,代码如下所示。

```xml
<?xml version="1.0" encoding="utf-8"?>
<LinearLayout xmlns:android="http://schemas.android.com/apk/res/android"
    android:orientation="vertical"
    android:layout_width="fill_parent"
    android:layout_height="fill_parent"
    >
    <ListView android:layout_height="wrap_content" android:id="@+id/listView1"
    android:layout_width="match_parent"></ListView>
</LinearLayout>
```

②新建一个 JavaBean：person.java。

```java
package com.xiao;
public class person {
    private Integer id;
    private String name;
    private Short age;
    public Integer getId() {
        return id;
    }
    public void setId(Integer id) {
        this.id = id;
    }
    public String getName() {
        return name;
    }
    public void setName(String name) {
        this.name = name;
    }
    public Short getAge() {
        return age;
    }
    public void setAge(Short age) {
        this.age = age;
    }
}
```

③新建 XmlContentHandler.java。

```java
package com.xiao;
import java.util.ArrayList;
import java.util.List;
import org.xml.sax.Attributes;
import org.xml.sax.SAXException;
import org.xml.sax.helpers.DefaultHandler;
public class XmlContentHandler extends DefaultHandler {
    private List<person> persons = null;
    private person currentPerson;
    private String tagName = null;//当前解析的元素标签
    public List<person> getPersons() {
        return persons;
    }
    /*
```

* 接收文档的开始的通知。
 */
@Override public void startDocument() throws SAXException {
        persons = new ArrayList<person>();
}
/*
 * 接收字符数据的通知。
 */
@Override public void characters(char[] ch, int start, int length) throws SAXException {
        if(tagName! = null) {
            String data = new String(ch, start, length);
            if(tagName.equals("name")) {
                this.currentPerson.setName(data);
            } else if(tagName.equals("age")) {
                this.currentPerson.setAge(Short.parseShort(data));
            }
        }
}
/*
 * 接收元素开始的通知。
 * 参数意义如下：
 *   namespaceURI：元素的命名空间
 *   localName：元素的本地名称（不带前缀）
 *   qName：元素的限定名（带前缀）
 *   atts：元素的属性集合
 */
@Override public void startElement(String namespaceURI, String localName, String qName, Attributes atts) throws SAXException {
        if(localName.equals("person")) {
            currentPerson = new person();
            currentPerson.setId(Integer.parseInt(atts.getValue("id")));
        }
        this.tagName = localName;
}
/*
 * 接收文档的结尾的通知。
 * 参数意义如下：
 * uri：元素的命名空间
 * localName：元素的本地名称（不带前缀）
 * name：元素的限定名（带前缀）

```
    *
    */
    @Override public void endElement(String uri, String localName, String name) throws SAXException {
        if(localName.equals("person")) {
            persons.add(currentPerson);
            currentPerson = null;
        }
        this.tagName = null;
    }
}
```

④新建 Activity：XmlTestActivity.java。

```
package com.xiao;
import java.io.File;
import java.io.FileInputStream;
import java.io.FileNotFoundException;
import java.io.IOException;
import java.util.ArrayList;
import java.util.HashMap;
import java.util.List;
import javax.xml.parsers.ParserConfigurationException;
import javax.xml.parsers.SAXParser;
import javax.xml.parsers.SAXParserFactory;
import org.xml.sax.InputSource;
import org.xml.sax.SAXException;
import org.xml.sax.XMLReader;
import android.app.Activity;
import android.os.Bundle;
import android.util.Log;
import android.widget.ListView;
import android.widget.SimpleAdapter;
public class XmlTestActivity extends Activity {
    /** Called when the activity is first created. */
    ListView l;
    @Override
    public void onCreate(Bundle savedInstanceState) {
        super.onCreate(savedInstanceState);
        setContentView(R.layout.main);
        l = (ListView)findViewById(R.id.listView1);
        SAXParserFactory spf = SAXParserFactory.newInstance();
        try {
```

```java
                SAXParser sp = spf.newSAXParser();
                XMLReader xr = sp.getXMLReader();
                XmlContentHandler myHandler = new XmlContentHandler();
                xr.setContentHandler(myHandler);
                xr.parse(new InputSource(new FileInputStream(new File("/sdcard/person.xml"))));
                List<person> temp = myHandler.getPersons();
                List<HashMap<String,String>> data = new ArrayList<HashMap<String,String>>();
                Log.v("person size -- >", temp.size()+"");
                for(person f:temp){
                    HashMap<String,String> person = new HashMap<String,String>();
                    person.put("name", f.getName());
                    person.put("age", f.getAge()+"");
                    data.add(person);
                }
                SimpleAdapter sa = new SimpleAdapter(XmlTestActivity.this, data, android.R.layout.simple_list_item_2, new String[]{"name","age"}, new int[]{android.R.id.text1, android.R.id.text2});
                l.setAdapter(sa);
            } catch (ParserConfigurationException e) {
                // TODO Auto-generated catch block
                e.printStackTrace();
            } catch (SAXException e) {
                // TODO Auto-generated catch block
                e.printStackTrace();
            } catch (FileNotFoundException e) {
                // TODO Auto-generated catch block
                e.printStackTrace();
            } catch (IOException e) {
                // TODO Auto-generated catch block
                e.printStackTrace();
            }
        }
    }
}
```

⑤新建 XMLTestActivity.java。

```java
package com.xiao;
import java.io.File;
import java.io.FileInputStream;
import java.io.FileNotFoundException;
import java.io.IOException;
import java.util.List;
import javax.xml.parsers.ParserConfigurationException;
```

```java
import javax.xml.parsers.SAXParser;
import javax.xml.parsers.SAXParserFactory;
import org.xml.sax.InputSource;
import org.xml.sax.SAXException;
import org.xml.sax.XMLReader;
import android.app.Activity;
import android.os.Bundle;
import android.util.Log;
import android.widget.ArrayAdapter;
import android.widget.ListView;
public class XmlTestActivity extends Activity {
    /** Called when the activity is first created. */
    ListView l;
    @Override
    public void onCreate(Bundle savedInstanceState) {
        super.onCreate(savedInstanceState);
        setContentView(R.layout.main);
        l = (ListView)findViewById(R.id.listView1);
        //SAX 解析例子
        SAXParserFactory spf = SAXParserFactory.newInstance();
        try {
            SAXParser sp = spf.newSAXParser();
            XMLReader xr = sp.getXMLReader();
            XmlContentHandler myHandler = new XmlContentHandler();
            xr.setContentHandler(myHandler);
            xr.parse(new InputSource(new FileInputStream(new File("/sdcard/person.xml"))));
            List<person> temp = myHandler.getPersons();
            Log.v("person size -->", temp.size()+"");
            for(person f:temp) {
                Log.v("name", f.getName());
                Log.v("age", f.getAge()+"");
            }
        } catch (ParserConfigurationException e) {
            // TODO Auto-generated catch block
            e.printStackTrace();
        } catch (SAXException e) {
            // TODO Auto-generated catch block
            e.printStackTrace();
        } catch (FileNotFoundException e) {
            // TODO Auto-generated catch block
```

```
            e. printStackTrace();
        } catch (IOException e) {
            // TODO Auto-generated catch block
            e. printStackTrace();
        }
    }
}
```

(2) 代码分析

XMLReader xr = sp. getXMLReader();

XmlContentHandler myHandler = new XmlContentHandler();

xr. setContentHandler(myHandler);

xr. parce(new InputSource(new FileInputStream(new File("/sdcard/person. xml"))));

其中获得 XMLReader 的固定步骤，设置了 myHandler 作为 xr 的读数据流的事件处理器，当特定的事件产生后，就会激活 myHandler 中固定的方法。

2) XML **数据解析**——DOM

除了可以使用 SAX 解析 XML 文件，也可以使用 DOM 解析 XML 文件。DOM 解析 XML 文件时，会将 XML 文件的所有内容读取到内存中，然后允许用户使用 DOM API 遍历 XML 树并检索所需的数据。使用 DOM 操作 XML 的代码看起来比较直观，并且在某些方面比基于 SAX 的实现更加简单。但是，因为 DOM 需要将 XML 文件的所有内容读取到内存中，所以内存的消耗比较大，特别对于运行 Android 的移动设备来说，因为设备的资源比较宝贵，所以建议还是采用 SAX 解析 XML 文件，当然，如果 XML 文件的内容比较少，采用 DOM 是可行的。

【**例** 5-2】 通过 DOM，实现与例 5-1 相同的功能。

(1) 实现过程

①参照项目 1，在 Eclipse 中创建名为 ex05_2DOM 的工程。person. java 和 Acitvity 的建立与上节相同。

②新建 DomXMLReader. java。

```
package com. xiao;
import java. io. InputStream;
import java. util. ArrayList;
import java. util. List;
import javax. xml. parsers. DocumentBuilder;
import javax. xml. parsers. DocumentBuilderFactory;
import org. w3c. dom. Document;
import org. w3c. dom. Element;
import org. w3c. dom. Node;
import org. w3c. dom. NodeList;
/ **
 * 使用 Dom 解析 XML 文件
 *
```

```java
*/
public class DomXMLReader {
    public static List<person> readXML(InputStream inStream) {
        List<person> persons = new ArrayList<person>();
        DocumentBuilderFactory factory = DocumentBuilderFactory.newInstance();
        try {
            DocumentBuilder builder = factory.newDocumentBuilder();
            Document dom = builder.parse(inStream);
            Element root = dom.getDocumentElement();
            NodeList items = root.getElementsByTagName("person");//查找所有person节点
            for (int i=0; i < items.getLength(); i++) {
                person person = new person();
                //得到第一个person节点
                Element personNode = (Element) items.item(i);
                //获取person节点的id属性值
                person.setId(new Integer(personNode.getAttribute("id")));
                //获取person节点下的所有子节点(标签之间的空白节点和name/age元素)
                NodeList childsNodes = personNode.getChildNodes();
                for (int j=0; j < childsNodes.getLength(); j++) {
                    Node node = (Node) childsNodes.item(j);//判断是否为元素类型
                    if(node.getNodeType() == Node.ELEMENT_NODE) {
                        Element childNode = (Element) node;
                        //判断是否为name元素
                        if ("name".equals(childNode.getNodeName())) {
                            //获取name元素下的Text节点,然后从Text节点获取数据
                            person.setName(childNode.getFirstChild().getNodeValue());
                        }
                        else if ("age".equals(childNode.getNodeName())) {
                            person.setAge(new Short(childNode.getFirstChild().getNodeValue()));
                        }
                    }
                }
                persons.add(person);
            }
            inStream.close();
        } catch (Exception e) {
            e.printStackTrace();
        }
        return persons;
    }
```

}

替换上面 Activity 中的解析方式。

```
List < person >  temp = null;
    try{
    temp = DomXMLReader. readXML( new FileInputStream( new File("/sdcard/person. xml")));
    } catch (FileNotFoundException e) {
        // TODO Auto-generated catch block
        e. printStackTrace();
    }
    for( person f:temp) {
        HashMap < String, String > person = new HashMap < String,String > ();
        person. put("name", f. getName());
        person. put("age", f. getAge() + "");
        data. add(person);
    }
    SimpleAdapter sa = new SimpleAdapter( XmlTestActivity. this, data, android. R. layout. simple_list_item_2,new String[ ]{"name","age"},new int[ ]{android. R. id. text1,android. R. id. text2});
    l. setAdapter(sa);
```

（2）代码分析

①产生解析器。

```
List < person >  temp = null;
    try{
    temp = DomXMLReader. readXML( new FileInputStream( new File("/sdcard/person. xml")));
```

②DOM 的解析，其实就是通过 XML 产生文档结构树，然后通过遍历节点方式，访问所需要的数据。

3）XML 数据解析——PULL

PULL 是一种基于事件的解析 XML 文件的方式。PULL 解析 XML 文件同 SAX 和 DOM 一样，都可以脱离 Android 单独使用，不同的是 PULL 读取 XML 文件后调用方法返回的是数字。

读取到 XML 的声明返回数字 0(START_DOCUMENT)；

读取到 XML 的结束返回数字 1(END_DOCUMENT)；

读取到 XML 的开始标签返回数字 2(START_TAG)；

读取到 XML 的结束标签返回数字 3(END_TAG)；

读取到 XML 的文本返回数字 4(TEXT)。

其中：

XmlPullParser. START_DOCUMENT 表示开始文档事件；

XmlPullParser. START_TAG:开始标签；

XmlPullParser. END_TAG:结束标签；

parser. getName():获取节点的名称；

parser. nextText():获取下一个 text 类型的节点；

parser. getAttributeValue(0)):获取属性值;

event = parser. next():继续下一个元素。

【例 5-3】 通过 PULL,实现与例 5-1 相同的功能。

(1) 实现过程

①使用上节创建好的工程。

其中,person. java 和 Acitvity 的建立与上节相同。

②新建 PullParseService. java。

```
package com. xiao;
import java. io. InputStream;
import java. util. ArrayList;
import java. util. List;
import org. xmlpull. v1. XmlPullParser;
import android. util. Xml;
public class PullParseService {
public List < person > parseDateSource(InputStream inputStream) throws Exception
{ List < person > persons = null;
    person currentPerson = null;
    XmlPullParser parse = Xml. newPullParser();
    parse. setInput(inputStream, "utf - 8");
    int event = parse. getEventType();
    // Returns the type of the current event (START_TAG, END_TAG, TEXT, etc.)
    while(event! = XmlPullParser. END_DOCUMENT) {
    switch (event) {
    case XmlPullParser. START_DOCUMENT:
            persons = new ArrayList < person > ();//初始化 books 集合
            break;
    case XmlPullParser. START_TAG:
                if(parse. getName(). equals("person")) {
                    currentPerson = new person();
                    //book. setId(Integer. parseInt(parse. getAttributeValue(0)));
                    //或者这样也可以的
                    currentPerson. setId(Integer. parseInt(parse. getAttributeValue(null, "id")));
                }
                if(currentPerson! = null) {
                    if(parse. getName(). equals("name")) {
                        currentPerson. setName(parse. nextText());
                    } else if(parse. getName(). equals("age")) {
                        currentPerson. setAge(Short. parseShort(parse. nextText()));
                    }
```

```
            }
            break;
        case XmlPullParser.END_TAG:
            if(parse.getName().equals("person")){
                persons.add(currentPerson);
                currentPerson = null;
            }
            break;
        default:
            break;
        }
        event = parse.next();//进入到下一个元素并触发相应事件
    }
    }
    return persons;
    }
}
```

③替换主 Activity 中的解析方式。
```
List<person> temp = null;
PullParseService pullParseService = new PullParseService();
try{
  temp = pullParseService.parseDateSource(new FileInputStream(new File("/sdcard/person.xml")));
} catch (Exception e) {
    e.printStackTrace();
}
for(person f:temp){
        HashMap<String,String> person = new HashMap<String,String>();
        person.put("name", f.getName());
        person.put("age", f.getAge()+"");
        data.add(person);
}
SimpleAdapter sa = new SimpleAdapter(XmlTestActivity.this, data, android.R.layout.simple_list_item_2,
new String[]{"name","age"}, new int[]{android.R.id.text1,android.R.id.text2});
l.setAdapter(sa);
```

（2）代码分析

上面 PULL 的解析方式中，通过一个 while 循环，在读取 XML 文件时，不断产生不同的事件，while 循环体针对不同的事件类型分别进行相应的处理。

3. 网络访问类——HttpClient 类

HttpClient 程序包是一个实现了 HTTP 协议的客户端编程工具包，要想熟练地掌握它，必须熟悉 HTTP 协议。

4. Android 多线程机制

1）Android 的单线程模型

当一个程序第一次启动时，Android 会同时启动一个对应的主线程（Main Thread）。主线程主要负责处理与 UI 相关的事件，如用户的按键事件、用户接触屏幕的事件以及屏幕绘图事件，并把相关的事件分发到对应的组件进行处理。所以，主线程通常又被叫做 UI 线程。在开发 Android 应用时必须遵守单线程模型的原则：Android UI 操作并不是线程安全的，并且这些操作必须在 UI 线程中执行。

在没有理解单线程模型的情况下，设计的程序可能会性能低下，因为所有的动作都在同一个线程中触发。例如，当主线程正在做一些比较耗时的操作时，如正从网络上下载一个大图片，或者访问数据库，由于主线程被这些耗时的操作阻塞住，无法及时响应用户的事件，从用户的角度看会觉得程序已经死掉。如果程序长时间不响应，用户还可能得重启系统。为了避免这样的情况，Android 设置了一个 5 s 的超时时间。一旦用户的事件由于主线程阻塞而超过 5 s 没有响应，Android 会弹出一个应用程序没有响应的对话框。

要演示超时的现象，只需要制造一种网络异常的状况。最简单的方式是断开网络连接，然后启动该程序，同时触发一个用户事件，如按一下 MENU 键，由于主线程因为网络异常而被长时间阻塞，所以用户的按键事件在 5 s 内得不到响应，Android 会提示一个程序无法响应的异常。该对话框会询问用户是继续等待还是强行退出程序。当程序需要去访问未知的网络时都可能发生类似的超时情况，用户的响应得不到及时回应会大大降低用户体验。所以，需要尝试以别的方式来实现。

2）子线程更新 UI

显然，如果程序需要执行耗时的操作，像上例一样由主线程来负责执行该操作是错误的。所以，需要创建一个新的子线程来负责调用天气预报 API 来获得天气数据。刚接触 Android 的开发者最容易想到的方式就是如下：

```
public void onClick( View v) {
    //创建一个子线程执行耗时的从网络上获取天气信息的操作
    new Thread( ) {
      @Override
      public void run( ) {
        //获得用户输入的城市名称
        String city = editText. getText( ). toString( ) ;
        //调用 Google 天气 API 查询指定城市的当日天气情况
        String weather = getWetherByCity( city) ;
        //把天气信息显示在 title 上
        setTitle( weather) ;
      }
    }. start( ) ;
}
```

但是，会发现 Android 提示程序由于异常而终止。为什么在其他平台上看起来很简单的代码在 Android 上运行的时候会出错呢？如果观察 LogCat 中打印的日志信息就会发现如下的错误日志：

android. view. ViewRoot $ CalledFromWrongThreadException：Only the original thread that created a view hierarchy can touch its views.

从错误信息不难看出，Android 禁止其他子线程来更新由 UI thread 创建的试图。本例中显示天气信息实际是一个由 UI thread 创建的 TextView，所以尝试在一个子线程中更改 TextView 的时候就出错了。这显示违背了单线程模型的原则：Android UI 操作并不是线程安全的，并且这些操作必须在 UI 线程中执行。

3) Message Queue

在单线程模型下，为了解决类似的问题，Android 设计了一个 Message Queue（消息队列），线程间可以通过该 Message Queue 并结合 Handler 和 Looper 组件进行信息交换。

Message Qucuo 是一个消息队列，用来存放通过 Handler 发布的消息。消息队列通常附属于某一个创建它的线程，可以通过 Looper. myQueue( ) 得到当前线程的消息队列。Android 在第一程序启动时会默认为 UI thread 创建一个关联的消息队列，用来管理程序的一些上层组件，如 activities、broadcast receivers 等。可以在自己的子线程中创建 Handler 与 UI thread 通信。

4) Handler

Handler 主要接收子线程发送的数据，并用此数据配合主线程更新 UI。

(1) android 的事件处理机制

当应用程序启动时，Android 首先会开启一个主线程（也就是 UI 线程），主线程为管理界面中的 UI 控件，进行事件分发。例如，如果单击一个 Button，Android 会分发事件到 Button 上来响应操作。如果此时需要一个耗时的操作，如联网读取数据，或者读取本地较大的一个文件，则不能把这些操作放在主线程中。这个时候需要把这些耗时的操作放在一个子线程中。因为子线程涉及 UI 更新，Android 主线程是线程不安全的，也就是说，更新 UI 只能在主线程中更新，子线程中操作是危险的，Handler 可以解决这个复杂的问题。由于 Handler 运行在主线程中（UI 线程中），它与子线程可以通过 Message 对象传递数据，Handler 承担着接收子线程传过来的（子线程用 sedMessage( )方法传递）Message 对象（里面包含数据），把这些消息放入主线程队列中，配合主线程进行 UI 更新。

通过 Handler 可以发布或者处理一个消息或者 Runnable 的实例。每个 Handler 都会与唯一的一个线程以及该线程的消息队列管理关联。当创建一个新的 Handler 时，默认情况下，它将关联到创建它的这个线程和该线程的消息队列。也就是说，如果通过 Handler 发布消息，消息将只会发送到与它关联的这个消息队列，当然也只能处理该消息队列中的消息。

(2) Handler 的特点

handler 可以分发 Message 对象和 Runnable 对象到主线程中，每个 Handler 实例都会绑定到创建它的线程中（一般是位于主线程）。它有两个作用：

①安排消息或 Runnable 在某个主线程中执行。

②安排一个动作在不同的线程中执行。

Handler 中分发消息的主要方法：

- post(Runnable);
- postAtTime(Runnable,long);
- postDelayed(Runnable long);
- sendEmptyMessage(int);
- sendMessage(Message);
- sendMessageAtTime(Message,long);
- sendMessageDelayed(Message,long)。

以上 post 类方法允许安排一个 Runnable 对象到主线程队列中。sendMessage 类方法允许安排一个带数据的 Message 对象到队列中，等待更新。

（3）Handler 实例

子类需要继承 Handler 类，并重写 handleMessage(Message msg) 方法，用于接收线程数据。

以下为一个实例，它实现的功能为通过线程修改界面 Button 的内容。

```
public class MyHandlerActivity extends Activity {
    Button button;
    MyHandler myHandler;
    protected void onCreate(Bundle savedInstanceState) {
        super.onCreate(savedInstanceState);
        setContentView(R.layout.handlertest);
        button = (Button) findViewById(R.id.button);
        myHandler = new MyHandler();
        //当创建一个新的 Handler 实例时,它会绑定到当前线程和消息的队列中,开始分发数据
        // Handler 有两个作用
        //(1)定时执行 Message 和 Runnalbe 对象
        //(2)让一个动作在不同的线程中执行
        //它用以下方法安排消息
        // post(Runnable)
        // postAtTime(Runnable,long)
        // postDelayed(Runnable,long)
        // sendEmptyMessage(int)
        //sendMessage(Message);
        //sendMessageAtTime(Message,long)
        //sendMessageDelayed(Message,long)
        //以上以 post 开头的方法允许用户处理 Runnable 对象
        //sendMessage() 允许用户处理 Message 对象(Message 里可以包含数据)
        MyThread m = new MyThread();
```

```java
        new Thread(m).start();
    }
    /**
     * 接收消息,处理消息,此 Handler 会与当前主线程一起运行
     */
    class MyHandler extends Handler {
        public MyHandler() {
        }
        public MyHandler(Looper L) {
            super(L);
        }
        //子类必须重写此方法,接收数据
        @Override
        public void handleMessage(Message msg) {
            // TODO Auto-generated method stub
            Log.d("MyHandler", "handleMessage......");
            super.handleMessage(msg);
            //此处可以更新 UI
            Bundle b = msg.getData();
            String color = b.getString("color");
            MyHandlerActivity.this.button.append(color);
        }
    }
    class MyThread implements Runnable {
        public void run() {
            try {
                Thread.sleep(10000);
            } catch (InterruptedException e) {
                // TODO Auto-generated catch block
                e.printStackTrace();
            }
            Log.d("thread......", "mThread........");
            Message msg = new Message();
            Bundle b = new Bundle();//存放数据
            b.putString("color", "我的");
            msg.setData(b);
            MyHandlerActivity.this.myHandler.sendMessage(msg);//向 Handler 发送消息,更新 UI
        }
    }
}
```

5）Looper

Looper 扮演着 Handler 和消息队列之间通信桥梁的角色。程序组件首先通过 Handler 把消息传递给 Looper，Looper 把消息放入队列。Looper 也把消息队列里的消息广播给所有的 Handler，Handler 接收到消息后调用 handleMessage 进行处理。

①可以通过 Looper 类的静态方法 Looper.myLooper( ) 得到当前线程的 Looper 实例，如果当前线程未关联 Looper 实例，该方法将返回空。

②可以通过静态方法 Looper.getMainLooper( ) 得到主线程的 Looper 实例。

任务1　实现天气预报

1. 任务说明

使用查询天气预报的功能，能够通过查询指定城市的天气，并显示在桌面上。

2. 实现过程

参照项目1，在 Eclipse 中创建名为 ex05_4 的工程。

①打开 res\\values\\strings.xml 文件，添加如下内容。

```
<?xml version="1.0" encoding="utf-8"?>
<resources>
    <string name="hello">Hello World, WeatherSearch!</string>
    <string name="app_name">天气查询</string>
    <string name="menu_weather">天气查询</string>
    <string name="wea_hint">请输入城市名称</string>
    <string name="no_city">系统中没有该城市</string>
    <string name="wea_searching">获取天气信息</string>
    <string name="tip">提示</string>
    <string name="getting">获取中,请稍后...</string>
    <string name="get_nothing">获取失败,请重试</string>
    <string name="xml_error">XML 解析出错</string>
    <string name="ok">确定</string>
</resources>
```

②在 res\\values 中新建 colors.xml，添加如下内容。

```
<?xml version="1.0" encoding="utf-8"?>
<resources>
    <color name="white">#FFFFFFFF</color>
    <color name="transparency">#00000000</color>
    <color name="title_bg">#FF9ed913</color>
    <color name="end_color">#A0cfef83</color>
</resources>
```

③在 res\\layout 目录中新建 main.xml，如图 5.4 所示。

项目 5　Android 网络通信——天气预报

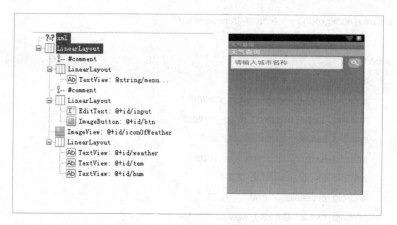

图 5.4　天气预报 XML 结构图

```xml
<?xml version="1.0" encoding="utf-8"?>
<LinearLayout xmlns:android="http://schemas.android.com/apk/res/android"
    android:orientation="vertical"
    android:layout_width="fill_parent"
    android:layout_height="fill_parent"
    android:background="@drawable/bg"
    >
    <!-- title 栏 -->
    <LinearLayout
        android:layout_width="fill_parent"
        android:layout_height="wrap_content"
        android:background="@color/title_bg"
        >
        <TextView
            android:layout_width="wrap_content"
            android:layout_height="wrap_content"
            android:layout_gravity="center_vertical"
            android:layout_marginLeft="5dp"
            android:textSize="18sp"
            android:textColor="@color/white"
            android:textStyle="bold"
            android:text="@string/menu_weather" />
    </LinearLayout>
    <!-- content 栏 -->
    <LinearLayout
        android:layout_width="fill_parent"
        android:layout_height="wrap_content"
        android:layout_marginTop="2dp"
```

```xml
        android:background = " @color/transparency "
        >
    <EditText
        android:id = " @ + id/input "
        android:layout_width = " wrap_content "
        android:layout_height = " wrap_content "
        android:layout_marginLeft = " 5dp "
        android:layout_marginRight = " 10dp "
        android:layout_weight = " 0.8 "
        android:clickable = " true "
        android:hint = " @string/wea_hint " />
    <ImageButton
        android:id = " @ + id/btn "
        android:layout_width = " wrap_content "
        android:layout_height = " wrap_content "
        android:layout_marginRight = " 5dp "
        android:layout_marginTop = " 5dp "
        android:background = " @drawable/search "
        android:clickable = " true "
        android:maxHeight = " 40dp "
        android:maxWidth = " 40dp "
        android:onClick = " onClick "
        android:src = " @drawable/search " />
</LinearLayout>
<ImageView
    android:id = " @ + id/iconOfWeather "
    android:layout_width = " wrap_content "
    android:layout_height = " wrap_content "
/>
<LinearLayout
    android:layout_width = " fill_parent "
    android:layout_height = " wrap_content "
    android:orientation = " vertical " >
    <TextView
        android:id = " @ + id/weather "
        android:layout_width = " wrap_content "
        android:layout_height = " wrap_content " />
    <TextView
        android:id = " @ + id/tem "
        android:layout_width = " wrap_content "
```

```
                    android:layout_height = " wrap_content " / >
                <TextView
                    android:id = " @ + id/hum "
                    android:layout_width = " wrap_content "
                    android:layout_height = " wrap_content " / >
            </LinearLayout>
</LinearLayout>
```

④新建 Activity：CurrentWeather.java，代码如下所示。

```java
package com.studio.android.chp07.ex04;
import java.io.BufferedInputStream;
import java.io.InputStream;
import java.io.UnsupportedEncodingException;
import java.net.URL;
import java.net.URLConnection;
import java.net.URLEncoder;
import javax.xml.parsers.SAXParser;
import javax.xml.parsers.SAXParserFactory;
import org.xml.sax.InputSource;
import org.xml.sax.XMLReader;
import android.app.Activity;
import android.graphics.Bitmap;
import android.graphics.BitmapFactory;
import android.os.Bundle;
import android.util.Log;
import android.view.View;
import android.view.View.OnClickListener;
import android.widget.Button;
import android.widget.EditText;
import android.widget.ImageView;
import android.widget.TextView;
public class CurrentWeather extends Activity {
    /** Called when the activity is first created. */
    @Override
    public void onCreate(Bundle savedInstanceState) {
        super.onCreate(savedInstanceState);
        setContentView(R.layout.main);
        Button submit = (Button) findViewById(R.id.btn);
        submit.setOnClickListener(new OnClickListener() {
            @Override
            public void onClick(View arg0) {
```

```java
try {
    String city = ( ( EditText) findViewById( R. id. input) )
        . getText( ). toString( ) ;
    try {
        city = URLEncoder. encode( city, "utf-8" ) ;
    } catch ( UnsupportedEncodingException e) {
        e. printStackTrace( ) ;
    }
    String queryString = " http://www. webxml. com. cn/WebServices/WeatherWebService. asmx/getWeatherbyCityName? theCityName = "
        + city;
    /* 将可能的空格替换为"%20" */
    URL aURL = new URL( queryString) ;
    /* 从 SAXParserFactory 获取 SAXParser */
    SAXParserFactory spf = SAXParserFactory. newInstance( ) ;
    SAXParser sp = spf. newSAXParser( ) ;
    //从 SAXParser 得到 XMLReader
    XMLReader xr = sp. getXMLReader( ) ;
    GoogleWeatherHandler gwh = new GoogleWeatherHandler( ) ;
    xr. setContentHandler( gwh) ;
    xr. parse( new InputSource( aURL. openStream( ) ) ) ;
    TextView tv1 = ( TextView) findViewById( R. id. tem) ;
    tv1. setText("温度:" + gwh. getTempData( ). get(5) + "摄氏度" ) ;
    TextView tv2 = ( TextView) findViewById( R. id. weather) ;
    tv2. setText( gwh. getTempData( ). get(6) ) ;
    TextView tv3 = ( TextView) findViewById( R. id. hum) ;
    tv3. setText(" " + gwh. getTempData( ). get(11) ) ;
    URL iconURL = new URL( " http://www. webxml. com. cn/images/weather/" + gwh. getTempData( ). get(8) ) ;
    URLConnection conn = iconURL. openConnection( ) ;
    conn. connect( ) ;
    InputStream is = conn. getInputStream( ) ;
    BufferedInputStream bis = new BufferedInputStream( is) ;
    //设置 icon
    ImageView iv = ( ImageView) findViewById( R. id. iconOfWeather) ;
    Bitmap bm = null;
    bm = BitmapFactory. decodeStream( bis) ;
    iv. setImageBitmap( bm) ;
    bis. close( ) ;
    is. close( ) ;
```

```
            } catch ( Exception e ) {
                Log. e( "error" ,e. toString( ) ) ;
            }
        }//end of onClick
    } ) ;//end of SetClick
    }
}
```

⑤新建 GoogleWeatherHandler. java，修改代码，如下所示。

```
import java. util. ArrayList;
import java. util. List;
import org. xml. sax. Attributes;
import org. xml. sax. SAXException;
import org. xml. sax. helpers. DefaultHandler;
/**
SAXHandler:用户从 Google Weather API 返回的 XML 中提取当前天气信息
*/
public class GoogleWeatherHandler extends DefaultHandler {
    ArrayList < String >    tempData;
    String TagName = " " ;
    public ArrayList < String >    getTempData( ) {
        return this. tempData;
    }
    @Override
    public void startDocument( ) throws SAXException {
        tempData = new ArrayList < String > ( ) ;
    }
    @Override
    public void endDocument( ) throws SAXException {
    }
    @Override
    public void startElement( String namespaceURI, String localName,
    String qName, Attributes atts) throws SAXException {
        // Outer´Tags
        if ( localName. equals( "string" ) ) {
            TagName = localName;
        }
    }
    @Override
    public void endElement( String namespaceURI, String localName, String qName)
        throws SAXException {
```

```java
            if(localName.equals("string")){
                TagName = "";
            }
        }
        @Override
        public void characters(char ch[], int start, int length) {
            if(TagName.equals("string")){
                String temp = new String(ch,start,length);
                tempData.add(temp);
            }
        }
    }
}
```

⑥修改上节中的 CurrentWeather.java 程序，其他与上节中例子相同。

```java
package com.studio.android.chp07.ex04;
import java.io.BufferedInputStream;
import java.io.BufferedReader;
import java.io.InputStream;
import java.io.InputStreamReader;
import java.io.UnsupportedEncodingException;
import java.net.URL;
import java.net.URLConnection;
import java.net.URLEncoder;
import javax.xml.parsers.SAXParser;
import javax.xml.parsers.SAXParserFactory;
import org.apache.http.HttpEntity;
import org.apache.http.HttpResponse;
import org.apache.http.HttpStatus;
import org.apache.http.client.HttpClient;
import org.apache.http.client.methods.HttpGet;
import org.apache.http.impl.client.DefaultHttpClient;
import org.apache.http.protocol.BasicHttpContext;
import org.apache.http.protocol.HttpContext;
import org.apache.http.util.EntityUtils;
import org.xml.sax.InputSource;
import org.xml.sax.XMLReader;
import android.app.Activity;
import android.graphics.Bitmap;
import android.graphics.BitmapFactory;
import android.os.Bundle;
import android.os.Handler;
```

```java
import android.os.Looper;
import android.os.Message;
import android.util.Log;
import android.view.View;
import android.view.View.OnClickListener;
import android.widget.Button;
import android.widget.EditText;
import android.widget.ImageButton;
import android.widget.ImageView;
import android.widget.TextView;
public class CurrentWeather_HttpClient extends Activity {
    /** Called when the activity is first created. */
    String queryString = "http://www.webxml.com.cn/WebServices/WeatherWebService.asmx/getWeatherbyCityName?theCityName=";
    URL aURL;
    @Override
    public void onCreate(Bundle savedInstanceState) {
        super.onCreate(savedInstanceState);
        setContentView(R.layout.main);
        ImageButton submit = (ImageButton) findViewById(R.id.btn);
        submit.setOnClickListener(new OnClickListener() {
            @Override
            public void onClick(View arg0) {
                /*获取用户输入的城市名称*/
                String city = ((EditText) findViewById(R.id.input)).getText().toString();
                try {
                    city = URLEncoder.encode(city, "utf-8");
                } catch (UnsupportedEncodingException e) {
                    e.printStackTrace();
                }
                getWeatherByCity(city);
                TextView tv1 = (TextView) findViewById(R.id.tem);
                tv1.setText("温度:" + gwh.getTempData().get(5) + "摄氏度");
                TextView tv2 = (TextView) findViewById(R.id.weather);
                tv2.setText(gwh.getTempData().get(6));
                TextView tv3 = (TextView) findViewById(R.id.hum);
                tv3.setText("" + gwh.getTempData().get(11));
                URL iconURL = new URL("http://www.webxml.com.cn/images/weather/" + gwh.getTempData().get(8));
                URLConnection conn = iconURL.openConnection();
```

```
        conn.connect();
    InputStream is = conn.getInputStream();
    BufferedInputStream bis = new BufferedInputStream(is);
    //设置icon
    ImageView iv = (ImageView)findViewById(R.id.iconOfWeather);
    Bitmap bm = null;
    bm = BitmapFactory.decodeStream(bis);
    iv.setImageBitmap(bm);
    bis.close();
    is.close();
}
public String getWeatherByCity(String city){
        HttpClient httpClient = new DefaultHttpClient();
        HttpContext localContext = new BasicHttpContext();
        HttpGet httpGet = new HttpGet(queryString + city);
        try{
            HttpResponse response = httpClient.execute(httpGet, localContext);
            if(response.getStatusLine().getStatusCode()!= HttpStatus.SC_OK){
                httpGet.abort();
            }else{
                HttpEntity httpEntitiy = response.getEntity();
                //解析字符串
                /* 从 SAXParserFactory 获取 SAXParser */
                SAXParserFactory spf = SAXParserFactory.newInstance();
                SAXParser sp = spf.newSAXParser();
                //从 SAXParser 得到 XMLReader
                XMLReader xr = sp.getXMLReader();
                GoogleWeatherHandler gwh = new GoogleWeatherHandler();
                xr.setContentHandler(gwh);
                xr.parse(new InputSource(response.getEntity().getContent()));
                return "天气情况";
            }
        }catch(Exception e){
        }finally{
            httpClient.getConnectionManager().shutdown();
        }
        return "网络异常";
    }
}
```

3. 代码分析

①http://www.webxml.com.cn/网站提供很多 Web Services，用于进行各种数据的 Web 服务。

http://www.webxml.com.cn/WebServices/WeatherWebService.asmx?op=getWeatherbyCityName 网站提供天气查询的 Web 服务，根据城市或地区名称查询获得未来三天内天气情况、现在的天气实况、天气和生活指数。调用方法如下：输入参数 theCityName = 城市中文名称（国外城市可用英文）或城市代码（不输入默认为上海市），如上海 或 58367，如有城市名称重复请使用城市代码查询（可通过 getSupportCity 或 getSupportDataSet 获得）；返回数据：一个一维数组 String(22)，共有 23 个元素。

String(0) 到 String(4) 分别代表省份、城市、城市代码、城市图片名称、最后更新时间。String(5) 到 String(11) 分别代表当天的 气温、概况、风向和风力、天气趋势开始图片名称（以下称图标一）、天气趋势结束图片名称（以下称图标二）、现在的天气实况、天气和生活指数。String(12) 到 String(16) 分别代表第二天的气温、概况、风向和风力、图标一、图标二。String(17) 到 String(21) 分别代表第三天的气温、概况、风向和风力、图标一、图标二。String(22) 代表被查询的城市或地区的介绍。

②Web 服务提供 SOAP1.1、SOAP1.2、GET 和 POST 不同方式进行查询。

- SOAP 1.1。

以下是 SOAP 1.1 请求和响应示例。所显示的占位符需替换为实际值。

POST /WebServices/WeatherWebService.asmx HTTP/1.1

Host：www.webxml.com.cn

Content-Type：text/xml；charset=utf-8

Content-Length：length

SOAPAction："http://WebXml.com.cn/getWeatherbyCityName"

<?xml version="1.0" encoding="utf-8"?>

<soap：Envelope xmlns：xsi="http://www.w3.org/2001/XMLSchema-instance" xmlns：xsd="http://www.w3.org/2001/XMLSchema" xmlns：soap="http://schemas.xmlsoap.org/soap/envelope/">

 <soap：Body>

  <getWeatherbyCityName xmlns="http://WebXml.com.cn/">

   <theCityName>string</theCityName>

  </getWeatherbyCityName>

 </soap：Body>

</soap：Envelope>

HTTP/1.1 200 OK

Content-Type：text/xml；charset=utf-8

Content-Length：length

<?xml version="1.0" encoding="utf-8"?>

<soap：Envelope xmlns：xsi="http://www.w3.org/2001/XMLSchema-instance" xmlns：xsd="http://www.w3.org/2001/XMLSchema" xmlns：soap="http://schemas.xmlsoap.org/soap/envelope/">

 <soap：Body>

```
        <getWeatherbyCityNameResponse xmlns="http://WebXml.com.cn/">
            <getWeatherbyCityNameResult>
                <string>string</string>
                <string>string</string>
            </getWeatherbyCityNameResult>
        </getWeatherbyCityNameResponse>
    </soap:Body>
</soap:Envelope>
```

- SOAP 1.2。

以下是 SOAP 1.2 请求和响应示例。所显示的占位符需替换为实际值。

```
POST /WebServices/WeatherWebService.asmx HTTP/1.1
Host: www.webxml.com.cn
Content-Type: application/soap+xml; charset=utf-8
Content-Length: length

<?xml version="1.0" encoding="utf-8"?>
<soap12:Envelope xmlns:xsi="http://www.w3.org/2001/XMLSchema-instance" xmlns:xsd="http://www.w3.org/2001/XMLSchema" xmlns:soap12="http://www.w3.org/2003/05/soap-envelope">
    <soap12:Body>
        <getWeatherbyCityName xmlns="http://WebXml.com.cn/">
            <theCityName>string</theCityName>
        </getWeatherbyCityName>
    </soap12:Body>
</soap12:Envelope>

HTTP/1.1 200 OK
Content-Type: application/soap+xml; charset=utf-8
Content-Length: length

<?xml version="1.0" encoding="utf-8"?>
<soap12:Envelope xmlns:xsi="http://www.w3.org/2001/XMLSchema-instance" xmlns:xsd="http://www.w3.org/2001/XMLSchema" xmlns:soap12="http://www.w3.org/2003/05/soap-envelope">
    <soap12:Body>
        <getWeatherbyCityNameResponse xmlns="http://WebXml.com.cn/">
            <getWeatherbyCityNameResult>
                <string>string</string>
                <string>string</string>
            </getWeatherbyCityNameResult>
        </getWeatherbyCityNameResponse>
    </soap12:Body>
</soap12:Envelope>
```

- HTTP GET。

以下是 HTTP GET 请求和响应示例。所显示的占位符需替换为实际值。

```
GET /WebServices/WeatherWebService.asmx/getWeatherbyCityName? theCityName = string HTTP/1.1
Host: www.webxml.com.cn
HTTP/1.1 200 OK
Content – Type: text/xml; charset = utf – 8
Content – Length: length
<? xml version = "1.0" encoding = "utf – 8"? >
< ArrayOfString xmlns = "http://WebXml.com.cn/" >
    < string > string </string >
    < string > string </string >
</ArrayOfString >
```

- HTTP POST。

以下是 HTTP POST 请求和响应示例。所显示的占位符需替换为实际值。

```
POST /WebServices/WeatherWebService.asmx/getWeatherbyCityName HTTP/1.1
Host: www.wobxml.com.cn
Content – Type: application/x – www – form – urlencoded
Content – Length: length
theCityName = string
HTTP/1.1 200 OK
Content – Type: text/xml; charset = utf – 8
Content – Length: length
<? xml version = "1.0" encoding = "utf – 8"? >
< ArrayOfString xmlns = "http://WebXml.com.cn/" >
    < string > string </string >
    < string > string </string >
</ArrayOfString >
```

- URL 类 get 方式访问某城市的天气预报。

```
SAXParserFactory spf = SAXParserFactory.newInstance();
SAXParser sp = spf.newSAXParser();
//从 SAXParser 得到 XMLReader
XMLReader xr = sp.getXMLReader();
GoogleWeatherHandler gwh = new GoogleWeatherHandler();
xr.setContentHandler(gwh);
xr.parse(new InputSource(aURL.openStream()));
```

提供 SAX 方式解析数据，xr 解析 URL 对应的数据流，边读边解析，解析过程中会产生不同的事件，事件的触发通过 GoogleWeatherHandler 中的不同方法触发。

上例中使用 HttpClient 的缺省实现类 DefaultHttpClient 实现了 Get 请求 Web 服务的数据。

```
HttpClient httpClient = new DefaultHttpClient();
    HttpContext localContext = new BasicHttpContext();
    HttpGet httpGet = new HttpGet(queryString + city);
```

```
try{
    HttpResponse response = httpClient.execute(httpGet, localContext);
    if(response.getStatusLine().getStatusCode()! = HttpStatus.SC_OK){
        httpGet.abort();
    }else{
        HttpEntity httpEntitiy = response.getEntity();
```

通过构建 HttpGet 构建 Get 请求，通过 HttpClient 对象发送请求，通过 HttpResponse 对象获取返回的数据。

任务2 实现天气预报多线程

1. 任务说明

修改任务1中的天气预报系统，使用多线程机制完善该天气预报系统。

2. 实现过程

修改 CurrentWeather_HttpClient.java。

1. package com.studio.android.chp07.ex04;
2. import java.io.BufferedInputStream;
3. import java.io.BufferedReader;
4. import java.io.InputStream;
5. import java.io.InputStreamReader;
6. import java.io.UnsupportedEncodingException;
7. import java.net.URL;
8. import java.net.URLConnection;
9. import java.net.URLEncoder;
10. import javax.xml.parsers.SAXParser;
11. import javax.xml.parsers.SAXParserFactory;
12. import org.apache.http.HttpEntity;
13. import org.apache.http.HttpResponse;
14. import org.apache.http.HttpStatus;
15. import org.apache.http.client.HttpClient;
16. import org.apache.http.client.methods.HttpGet;
17. import org.apache.http.impl.client.DefaultHttpClient;
18. import org.apache.http.protocol.BasicHttpContext;
19. import org.apache.http.protocol.HttpContext;
20. import org.apache.http.util.EntityUtils;
21. import org.xml.sax.InputSource;
22. import org.xml.sax.XMLReader;
23. import android.app.Activity;

```
24.    import android. graphics. Bitmap;
25.    import android. graphics. BitmapFactory;
26.    import android. os. Bundle;
27.    import android. os. Handler;
28.    import android. os. Looper;
29.    import android. os. Message;
30.    import android. util. Log;
31.    import android. view. View;
32.    import android. view. View. OnClickListener;
33.    import android. widget. Button;
34.    import android. widget. EditText;
35.    import android. widget. ImageButton;
36.    import android. widget. ImageView;
37.    import android. widget. TextView;
38.    public class CurrentWeather_HttpClient extends Activity {
39.        / * *  Called when the activity is first created. */
40.        String queryString = " http://www. webxml. com. cn/WebServices/WeatherWebService. asmx/getWeatherbyCityName? theCityName = ";
41.        MessageHandler messageHandler;
42.        URL aURL;
43.        @Override
44.        public void onCreate( Bundle savedInstanceState) {
45.            super. onCreate( savedInstanceState);
46.            setContentView( R. layout. main);
47.            ImageButton submit = ( ImageButton) findViewById( R. id. btn);
48.            Looper looper = Looper. myLooper( );
49.            messageHandler = new MessageHandler( looper);
50.            submit. setOnClickListener( new OnClickListener( ) {
51.              @Override
52.              public void onClick( View arg0) {
53.                /* 从 google 上获得图标 */
54.                new Thread( ) {
55.                  public void run( ) {
56.                    try {
57.                      /* 获取用户输入的城市名称 */
58.                      String city = ( ( EditText) findViewById( R. id. input) )
59.                          . getText( ). toString( );
60.                      try {
61.                        city = URLEncoder. encode( city, "utf -8");
62.                      } catch ( UnsupportedEncodingException e) {
```

```
63.                              e.printStackTrace();
64.                         }
65.                         getWeatherByCity(city);
66.                    }catch(Exception e){
67.                         Log.e("error",e.toString());
68.                    }
69.                    finally{
70.                         if(aURL!=null)
71.                              aURL=null;
72.                    }
73.               }//end of onClick
74.          }.start();
75.     }
76. });//end of SetClick
77. }
78. class MessageHandler extends Handler{
79.     public  MessageHandler(Looper looper){
80.          super(looper);
81.     }
82.     public void handleMessage(Message msg){
83.          GoogleWeatherHandler gwh=(GoogleWeatherHandler)msg.obj;
84.          try{
85.               TextView tv1=(TextView)findViewById(R.id.tem);
86.               tv1.setText("温度:"+gwh.getTempData().get(5)+"摄氏度");
87.               TextView tv2=(TextView)findViewById(R.id.weather);
88.               tv2.setText(gwh.getTempData().get(6));
89.               TextView tv3=(TextView)findViewById(R.id.hum);
90.               tv3.setText(" "+gwh.getTempData().get(11));
91.               URL iconURL=new URL("http://www.webxml.com.cn/images/weather/"+
                    gwh.getTempData().get(8));
92.               URLConnection conn=iconURL.openConnection();
93.               conn.connect();
94.               InputStream is=conn.getInputStream();
95.               BufferedInputStream bis=new BufferedInputStream(is);
96.               //设置icon
97.               ImageView iv=(ImageView)findViewById(R.id.iconOfWeather);
98.               Bitmap bm=null;
99.               bm=BitmapFactory.decodeStream(bis);
100.              iv.setImageBitmap(bm);
101.              bis.close();
```

```
102.                    is.close();
103.                } catch(Exception e){
104.                }
105.        }
106.    }
107.    public String getWeatherByCity(String city){
108.        HttpClient httpClient = new DefaultHttpClient();
109.        HttpContext localContext = new BasicHttpContext();
110.        HttpGet httpGet = new HttpGet(queryString + city);
111.        try{
112.            HttpResponse response = httpClient.execute(httpGet, localContext);
113.            if(response.getStatusLine().getStatusCode()! = HttpStatus.SC_OK){
114.                httpGet.abort();
115.            } else {
116.                HttpEntity httpEntitiy = response.getEntity();
117.                //解析字符串
118.                /* 从 SAXParserFactory 获取 SAXParser */
119.                SAXParserFactory spf = SAXParserFactory.newInstance();
120.                SAXParser sp = spf.newSAXParser();
121.                //从 SAXParser 得到 XMLReader
122.                XMLReader xr = sp.getXMLReader();
123.                GoogleWeatherHandler gwh = new GoogleWeatherHandler();
124.                xr.setContentHandler(gwh);
125.                xr.parse(new InputSource(response.getEntity().getContent()));
126.                Message message = Message.obtain();
127.                message.obj = gwh;
128.                messageHandler.sendMessage(message);
129.            return "天气情况";
130.            }
131.        } catch(Exception e){
132.        } finally{
133.            httpClient.getConnectionManager().shutdown();
134.        }
135.        return "网络异常";
136.    }
137. }
```

3. 代码分析

①访问网络的方法 getWeatherByCity(),放到一个子线程的 run()方法中启动。

②MessageHandler 在子线程中发送 Message (包含获取到的天气信息),由 MessageHandler 中的 handleMessage()方法进行主界面的更新。

功能拓展

虽然借助消息队列已经可以较为完美地实现天气预报的功能，但是用户还需要自己管理子线程，尤其当需要有一些复杂的逻辑以及需要频繁更新 UI 的时候，这样的方式会使得代码难以阅读和理解。

幸运的是 Android 另外提供了一个工具类：AsyncTask。它使得 UI thread 的使用变得异常简单。它使创建需要与用户界面交互的长时间运行的任务变得更简单，不需要借助线程和 Handler 即可实现。

①子类化 AsyncTask。

②实现 AsyncTask 中定义的下面一个或几个方法。

- onPreExecute()，该方法将在执行实际的后台操作前被 UI thread 调用。可以在该方法中做一些准备工作，如在界面上显示一个进度条。
- doInBackground(Params...)，将在 onPreExecute() 方法执行后马上执行，该方法运行在后台线程中。这里将主要负责执行那些很耗时的后台计算工作。可以调用 publishProgress() 方法来更新实时的任务进度。该方法是抽象方法，子类必须实现。
- onProgressUpdate(Progress...)，在 publishProgress() 方法被调用后，UI thread 将调用这个方法从而在界面上展示任务的进展情况，例如，通过一个进度条进行展示。
- onPostExecute(Result)，在 doInBackground 执行完成后，onPostExecute() 方法将被 UI thread 调用，后台的计算结果将通过该方法传递到 UI thread。

③为了正确地使用 AsyncTask 类，以下是几条必须遵守的准则。

- Task 的实例必须在 UI thread 中创建。
- execute() 方法必须在 UI thread 中调用。
- 不要手动调用 onPreExecute()、onPostExecute(Result)、doInBackground(Params...)、onProgressUpdate(Progress...) 这几个方法。
- 该 task 只能被执行一次，否则多次调用时会出现异常。

下面通过 AsyncTask 并且严格遵守上面的 4 条准则来改写天气预报的例子。

```
package com.studio.android.chp07.ex04;
import java.io.BufferedInputStream;
import java.io.InputStream;
import java.io.UnsupportedEncodingException;
import java.net.URL;
import java.net.URLConnection;
import java.net.URLEncoder;
import javax.xml.parsers.SAXParser;
import javax.xml.parsers.SAXParserFactory;
import org.apache.http.HttpEntity;
import org.apache.http.HttpResponse;
import org.apache.http.HttpStatus;
```

```java
import org.apache.http.client.HttpClient;
import org.apache.http.client.methods.HttpGet;
import org.apache.http.impl.client.DefaultHttpClient;
import org.apache.http.protocol.BasicHttpContext;
import org.apache.http.protocol.HttpContext;
import org.xml.sax.InputSource;
import org.xml.sax.XMLReader;
import android.app.Activity;
import android.graphics.Bitmap;
import android.graphics.BitmapFactory;
import android.os.AsyncTask;
import android.os.Bundle;
import android.os.Handler;
import android.os.Looper;
import android.os.Message;
import android.util.Log;
import android.view.View;
import android.view.View.OnClickListener;
import android.widget.Button;
import android.widget.EditText;
import android.widget.ImageView;
import android.widget.TextView;
public class CurrentWeather_AsyncTask extends Activity {
    /** Called when the activity is first created. */
    String queryString = "http://www.webxml.com.cn/WebServices/WeatherWebService.asmx/getWeatherbyCityName? theCityName=";
    @Override
    public void onCreate(Bundle savedInstanceState) {
        super.onCreate(savedInstanceState);
        setContentView(R.layout.main);
        Button submit = (Button) findViewById(R.id.btn);
        submit.setOnClickListener(new OnClickListener() {
            @Override
            public void onClick(View arg0) {
                /*从google上获得图标*/
                String city = ((EditText) findViewById(R.id.input))
                    .getText().toString();
                new GetWeatherTask().execute(city);
            }
        });//end of SetClick
    }
```

```java
class GetWeatherTask extends AsyncTask<String,Integer,String>{
    GoogleWeatherHandler gwh = new GoogleWeatherHandler();
    @Override
    protected String doInBackground(String... params){
        // TODO Auto-generated method stub
        String city = params[0];
        try{
            try{
                city = URLEncoder.encode(city,"utf-8");
            }catch(UnsupportedEncodingException e){
                e.printStackTrace();
            }
            queryString += city;
            URL aURL = new URL(queryString.replace(" ","%20"));
            /* 从 SAXParserFactory 获取 SAXParser */
            SAXParserFactory spf = SAXParserFactory.newInstance();
            SAXParser sp = spf.newSAXParser();
            //从 SAXParser 得到 XMLReader
            XMLReader xr = sp.getXMLReader();
            xr.setContentHandler(gwh);
            xr.parse(new InputSource(aURL.openStream()));
        }catch(Exception e){
            Log.e("error",e.toString());
        }
        return null;
    }
    @Override
    protected void onPostExecute(String result){
        // TODO Auto-generated method stub
        try{
            TextView tv1 = (TextView)findViewById(R.id.tem);
            tv1.setText("温度:" + gwh.getTempData().get(5) + "摄氏度");
            TextView tv2 = (TextView)findViewById(R.id.weather);
            tv2.setText(gwh.getTempData().get(6));
            TextView tv3 = (TextView)findViewById(R.id.hum);
            tv3.setText(" " + gwh.getTempData().get(11));
            URL iconURL = new URL("http://www.webxml.com.cn/images/weather/" + gwh.getTempData().get(8));
            URLConnection conn = iconURL.openConnection();
            conn.connect();
            InputStream is = conn.getInputStream();
```

```
                BufferedInputStream bis = new BufferedInputStream(is);
                //设置icon
                ImageView iv = (ImageView)findViewById(R.id.iconOfWeather);
                Bitmap bm = null;
                bm = BitmapFactory.decodeStream(bis);
                iv.setImageBitmap(bm);
                bis.close();
                is.close();
            }catch(Exception e){
            }
        }
    }
}
```

注意代码 new GetWeatherTask().execute(city)。必须每次都重新创建一个新的 GetWeatherTask 来执行后台任务，否则 Android 会提示 a task can be executed only once 的错误信息。经过改写后的程序非常简洁，减少了代码量，大大增强了可读性和可维护性。因为负责更新 UI 的 onPostExecute() 方法是由 UI thread 调用，所以没有违背单线程模型的原则。良好的 AsyncTask 设计大大降低了犯错误的概率。

 实战演练

如图 5.5 所示，实现下面所示的全部功能。

图 5.5　快捷查询系统

提示：参照 http://www.webxml.com.cn 提供的 Web 服务接口，为各种 Web 服务的查询编写一个查询客户端。

参 考 文 献

[1] 杨谊. Android 移动应用开发 [M]. 北京：人民邮电出版社, 2017.

[2] 刘佰龙. Android 移动应用开发教程 [M]. 北京：机械工业出版社, 2017.

[3] 陈文, 郭依正. 深入理解 Android 网络编程：技术详解与最佳实践 [M]. 北京：机械工业出版社, 2013.